ARN?
NARN!

FACES AND VOICES
OF
ATLANTIC INSHORE FISHERS

Best wishes to Peter —
the name of a good
fisherman —
Tom Roach

Thomas E. Roach

University College of Cape Breton Press

The University College of Cape Breton Press acknowledges the support received for its publishing program from the Canada Council's Block Grants program.

Cover design by Julie Scriver, Goose Lane Editions
Design and layout by Gail Jones
Printed and bound by City Printers, Sydney, Nova Scotia

Canadian Cataloguing in Publication Data
      Arn? Narn!
ISBN 0-920336-47-7

1. Fishers -- Atlantic Provinces -- social conditions. 2. Fishers --Atlantic Prov-inces -- Economic aspects. 3. Atlantic cod fishing -- Social aspects --Atlantic Provinces. 4. Atlantic cod fishing -- Economic conditions -- Atlantic Provinces. 5. Fishery policy -- Social aspects -- Atlantic Provinces. 6. Fishery policy -- Economic aspects -- Atlantic Provinces. I. Title

HD8089.F652C2 1997   338.3'72753   C95-950182-X

University College of Cape Breton Press
Box 5300
Sydney, Nova Scotia
CANADA

Make and Break Harbour
© Stan Rogers
Used with permission
Fogarty's Cove Music, 23 Hillside Ave. S., Dundas, Ontario L9H 4H7

CONTENTS

ARN?
NARN!

It was a beautiful sunny afternoon in August, 1993, when I first heard the above words. I was doing some sketching on the wharf at Petty Harbour, Newfoundland. As a boat entered the harbour, a man next to me yelled to the fisher on board, "Arn?" The reply came back, "Narn." When I got back to the Bed and Breakfast where I was staying, I asked for an explanation. Arn, the question, meant "Are there any?" and narn, the reply, was in the negative. Did you catch anything? Nothing. I knew I had my title.

To Iris, Peggy, Tom, Ofelia and Goyo

Preface

What effect did the cod moratorium have on the fishers of Atlantic Canada? In 1993, I decided the best way to answer this question was to speak directly to some of the people involved. I was determined to put together a visual record by sketching and interviewing inshore fishers in each of the four Atlantic provinces. Basically I asked three questions: "What caused the cod crisis?"; "How has it affected you?", and "What changes would you recommend?" Sometimes I sketched directly on location, and at other times I took photographs and worked up the sketches later. I drew what I saw and I listened carefully to the answers. When all the interviews were assembled, I realized the need for some sort of glossary or primer, and added the opening chapter on the groundfish fishery. While I was reflecting on the interviews and thinking about a suitable ending, I listened again to the lyrics of Make and Break Harbour by the late Stan Rogers. Through the magic of his song, Stan has given a pithy and poignant summary of the cod crisis. The words of the song are included at the end of this book. I added them as a summing up of sorts and as a tribute to the inshore fishers of Atlantic Canada, a valiant group of men and women who form the very backbone of life in our coastal communities. May they - like "old nets hung to dry" - not be "blown away, lost and forgotten."

Thomas E. Roach

Acknowledgements

I gratefully acknowledge the assistance of the Centre for Regional Studies at St. Francis Xavier University in Antigonish, Nova Scotia. The centre is funded by the Social Sciences and Humanities Research Council's Aid to Small Universities Program. The funding enabled me to travel to Louisbourg, Nova Scotia and Petty Harbour, Newfoundland in the summer of 1993, and to the north shore of New Brunswick and the east coast of Prince Edward Island in the summer of 1994. The members of the Petty Harbour Fishermen's Producer Co-operative Society Ltd. were very helpful, especially Tom Best. Reg and Cora Best invited me over for a feed of fresh cod. It was "some good." Dr. Robert Hennessey, a college classmate who lives in Bathurst, gave me an invaluable lead for New Brunswick. Arthur Smith from Stonehaven spent several days driving me up and down the north shore and introducing me to the inshore fishers. He and his wife Melba were kind and hospitable. Sasha Moon Corkum, a former student of mine from Prince Edward Island, directed me to her grandfather Jack King in Georgetown, P.E.I. Jack in turn sent me to his two sons, Danny and Leslie. One lead led to another and soon the interviews were completed. The overview of the groundfish fishery could not have been written without the generous collaboration of Stuart Beaton, a local fisherman. I also acknowledge the assistance of Dr. Ken den Heyer, the Dean of Arts, who granted me a six month sabbatical leave to complete this project. Thanks to Kate, Sharon and Gillian for pinchhitting for me, to my colleagues who supported and encouraged me, and to Ivy Green for typing the manuscript. Thanks to Dr. Aloysius Balawyder for his advice and assistance, and to Ray MacDonald for recording some sea songs and other melodies, and for help with copyright. Last, but not least, I am indebted to J. Arthur LeBlanc for writing the Foreward. Arthur was involved in co-operative fisheries with United Maritime Fishermen for thirty years, and held the position of executive director of the Co-operative Secretariat within the federal government for a period of three years prior to his retirement.

Foreword

Until the last half of the 20th century, it was generally thought that the fishery resources were unlimited, especially for groundfish and pelagic species because of the vast area of oceans they ranged. It became evident earlier that the sedentary stocks, such as most shellfish, were more at risk, and measures were taken to protect them from overfishing and the resulting depletion by establishing conservation measures such as seasons, minimum sizes and protection of spawning stocks. Protection of these stocks was somewhat easier because they were mostly, if not all, under the jurisdiction of the coastal state.

The situation was quite different when it became evident that the groundfish and pelagic species were coming under excessive fishing pressures. Prior to 1977, the jurisdiction of the coastal state was limited to three miles. These species migrate outside of the three mile limit during extended periods of the year, some even spend most, if not all of the time beyond the said limit, and they were subjected to extreme fishing effort by foreign fleets as well as by most components of our Canadian fishing fleet as it modernized and adopted ever more sophisticated technologies.

Enactment of a 200 mile fishing zone in 1977 was not the solution it was hoped to be. Historic rights gave continued access to the fleets of certain countries, and others gained access because of unutilized and/or under utilized species. Rapid expansion of the Canadian offshore and mid-shore fleets, and of plant capacity to process their landings, soon replaced the reduced foreign presence and effort. Quotas, minimum fish sizes, boat length limitations and mesh sizes had little impact on the volume of landings, not because the resource was able to withstand the increased pressure but because of the effectiveness of the more refined fish-finding and catching technologies.

The depletion of stocks as a result of the increased fishing effort impacted initially and most noticeably on the inshore. Not only was there a drastic decline in volume, but there was also a very noticeable decline in fish size. The inshore fishermen were the first to raise the alarm that the stocks were in trouble, that the stock assessments had not been and were not accurate, and that the fishing effort and landings had to be curtailed drastically. Their pleas fell on deaf ears; theirs was a voice crying in the wilderness. Those with major investments in draggers and processing facilities, who could ill afford any reduction in volume, denied the obvious. Vessel crews and plant workers also denied the obvious because any significant reduction in landings impacted directly and proportionately on their income. Governments were slow to react because they had vested interests: their substantial investment in the industry as well as the impact that any reduction would have on employment. Those groupings, and not the inshore fishermen, had the majority of representatives on the decision-making councils.

By delaying action, they were hoping that the impending crisis would somehow resolve itself. Instead, it worsened to the point where there was no other option but to impose a moratorium. The worst fears of the inshore fishermen were vindicated.

Much ink has flowed in the interim, some pointing the finger of blame, some attempting to justify the decisions and actions taken, and some attempting to rationalize the "why" of it all. Indeed, it is possible that changing water temperatures, that a rapidly increasing seal population, that depletion of certain key components in the food chain may have been contributing factors. But there is no doubt that some stock management decisions, that highly effective fish-finding and non-selective harvesting technologies, and that the dumping of under-size, non-targeted fish at sea have also been contributing factors. The damage has been done. With time, hopefully, the cod resource will recover. Hard decisions must now be taken to assure that the same mistakes are not repeated.

In "Arn? Narn!", Thomas Roach has focused on the inshore sector of the industry to put what he qualifies as a "human face" on the cod crisis. He has interviewed inshore fishermen in communities within each of the Atlantic provinces, asking them three specific questions: "What caused the cod crisis? and "How has it affected you?", "What changes would you recommend?"

Some essential elements are in evidence throughout the interviews. As to cause, the emphasis was on the introduction and use of technologies that, although highly productive proved to be very destructive over time, especially bottom trawling and gillnetting, the former because of non-selectivity of catch and destruction of habitat, and the latter because of continued "ghost-fishing" after being lost or abandoned at sea.

The impact on the inshore fishermen and their communities has varied widely depending upon the diversity of fish stocks in their areas. There is no doubt that Newfoundland fishermen and communities were hardest hit because of their dependence on groundfish, and especially on cod. Without the various assistance programs, the impact would have been more disastrous, and it may yet be so should the assistance be discontinued before the cod fishery can recover. It is evident to them that the fishery of the future shall be significantly different to what it has been, that the numbers involved therein will be reduced. For many, the uncertainty of it all is indeed depressing!

There is unanimity that a new management strategy must be adopted. The technologies to be used must be more selective and less damaging to the habitat. They must be more cost effective, with emphasis on maximizing rewarding employment rather than on the use of costly technologies and gadgetry. And the fishing effort must respect the essential balance in the food chain so that it can effectively support the stocks which are the major contributors to our food requirements, to the economy of our fishing communities in particular, and to the economy of Atlantic Canada generally.

This book demonstrates how oral history can be both interesting and effective. It is a valuable contribution to a better understanding of the crisis in the fishing industry. There are numerous scientific studies in this area, but no study has portrayed the anguish and plight of inshore fishermen, both in drawings and in words, as well as this one. The author has refrained from putting his own interpretation on the contents of the interviews. He has even faithfully retained the local vocabulary, expressions and phraseology which adds colour and makes for interesting reading. His sketches of the individuals interviewed and of the background are true to life. They prompt us to ask ourselves what lies in store for them and for their way of life. For how long and for how many shall the answer to the question "Arn?" continue to be "Narn."

J. Arthur LeBlanc

CHAPTER
1
THE GROUNDFISH FISHERY

CHAPTER 1

THE GROUNDFISH FISHERY

Commercial fishing on Canada's Atlantic coast is conducted by thousands of fishermen operating with many varieties of gear and vessels. The recent collapse of codfish stocks has put the groundfish sector of the fishery in the news. Public knowledge of the groundfish fishery is limited by snippets of information and incomplete news coverage on radio and television. This is not too surprising because reporters often have little in-depth experience of what is a rather complex activity. Moreover the terminology is often confusing and sometimes contradictory, e.g. in some areas, trawling refers to boats using hooks and line whereas in others it refers to the towing of nets.

After I had completed interviews with fishers and fish plant workers at four sites in the Atlantic region, I realized that some form of groundfish primer would help readers understand their comments. The Communications Branch of Fisheries and Oceans in Moncton, New Brunswick, supplied me with information on vessels and gear. Stuart Beaton, an experienced fisherman from Antigonish, supplied me with most of the material for the following overview.

GROUNDFISH

Groundfish live on or near the bottom of the ocean and include such species as cod, haddock and various types of sole and flounder. Fish, like birds, have migratory patterns. At certain times they may congregate close to shore, and at other times they may move to deeper water. Scientists have successfully traced the patterns of some groups or stocks, but others are less understood despite the best efforts by scientists.

FISHING FLEETS

Fishers generally fall into three categories depending on the size of the boats in which they work: offshore, mid-shore and inshore. Offshore boats are over 100 feet in length and are generally, but not always, owned by a company. Mid-shore fishers work in boats over 45 feet in length which are generally, but not always, owned by individuals. Inshore fishers work in boats under 45 feet in length and these boats are almost exclusively owned by individual fishers. In actual fishing operations, the activities of the various fleets overlap considerably because all three

3

fleets are after the same species — groundfish. Although some areas are exclusive to certain fleets, e.g. the extreme northern winter fishery to the offshore and the Northumberland Strait to the inshore, it is no more unusual to see 150 foot "offshore" boats fishing 12 to 15 miles off the west coast of Newfoundland than it is to see "inshore" boats fishing 120 miles from shore on Georges Bank.

GEAR TYPES

A further distinction between fleets and fishers is based on gear types. Gear is either passive or active. In passive gear types, the fish come to the gear, and in the active types, the gear moves over or around the fish. Fishers describe the two gear types as fixed gear and mobile gear.

QUOTA OR TOTAL ALLOWABLE CATCH

The Total Allowable Catch is the best estimate of the actual number of tons of fish which can be caught in a given fish stock in a particular year. This quota is apportioned to the fleets more or less as follows: offshore mobile gear, so many tons, offshore fixed gear, so many tons, mid-shore mobile gear, so many tons, and so on through each fleet division. This allocation is not equal but is based on a variety of factors such as the number of vessels in the fleet, its past history and the reliance by the particular fleet on the resource.

FIXED GEAR

Fixed gear is stationary on the bottom of the ocean while it is actually fishing. Gillnets have weights on the bottom and cork on the top so that they stand in an upright position on the seabed. Fish are caught as they attempt to swim through the webbing, entangling their gills. No bait is used. Buoys indicate the location and ownership of the gear and provide a line from which the gear can be raised to the surface. The nets may be moved from place to place by the fishers, and positioned in varying water depths, depending on the location of the species. Mesh size may differ according to the species and size of the fish sought.

The other major type of fixed gear is longline or tub trawl. Longline is just that—a long line with many baited hooks spread along the ocean floor. This system has now become mechanized with the use of automatic hauling, baiting and shooting machines. These improvements allow fishers to fish more gear and compete with other forms of fishing. This gear may also be moved from place to place, but it is stationary while it is actually working.

Cod trap fishing is primarily used in Newfoundland. The traps are open topped box nets with a vertical opening or door on one side. The trap is set on the ocean bottom, usually close to the shore, with the door facing shallow water. It is buoyed on the top and anchored at each corner. A long leader net extends from shallow water into the trap. When the cod meet the leader, they instinctively change direction, swimming through the open doors into the trap. Once inside, they tend to swim in circles in trying to avoid the leader and fail to locate the doors. When the doors are closed and the trap is hauled, the fishermen collect the fish with a dip net.

MOBILE GEAR

Mobile gear is not stationary while it is actually working. In general terms, boats using mobile gear are referred to as draggers or trawlers. In dragging, a cone-shaped net is lowered to the ocean floor by means of wire cables and towed along the ocean bottom to catch many species of groundfish. The net is kept open at the front end by large steel or weighted wooden "otterboards" called doors that are attached to cables between the boat and the net. The net traps fish in the end of the bag-like section or "cod-end," which has a mesh size that permits only the smaller fish to escape. The net and the doors maintain contact with the sea floor. The size, width and length of the net being used are determined by the size and towing power of the vessel. The width of the net may be from 70 feet to 250 feet. Tows may last for ten minutes or even several hours, and the distance traversed could be as much as thirty miles or more. After a period of towing, the trawl is winched up beside the vessel. In a stern trawling operation the gear is hauled up the ramp and the cod-end opened. Everything is retrieved on board the vessel, and no gear is left unattended on the bottom.

Although different gear types are found on vessels varying widely in size, in general, the vast majority of vessels over 100 feet are draggers, a majority of vessels 45 to 65 feet are draggers, and the vast majority of longliners and gillnetters are under 45 feet. Thus fixed gear is a predominantly inshore activity, and dragging is mainly a mid-shore and offshore activity.

SEINING

In purse seining, a wall of webbing is used to encircle the fish. A small boat pulls one end of the seine, and the vessel and boat encircle the fish with the net. The net is kept vertical in the water by floats on the top and weights on the bottom. A purse seine also has a wire rope passing through rings on the bottom of the net which enables the net to be drawn together to entrap the fish.

Danish and Scottish seining methods use similar nets and series of ropes spread out in a pear-shaped form along the ocean floor. Whereas, in Danish seining, the vessel remains in a fixed position while the gear is being hauled along the bottom. In Scottish seining, the net and ropes are towed along the ocean floor while they are closing.

Canadian pair seining uses two vessels in the operation. The net resembles an otter trawl net except that the vertical opening is much wider. When the two vessels come together, the cables are brought together and the net is winched in from both boats.

Mid-water trawls resemble otter trawls in that they are cone shaped and constructed of webbing. However they have fewer weights and can be adjusted for towing at various depths.

On side trawlers, the gear is towed from gallows fixed on one side of the vessel. These vessels are between 65 to 100 feet in length and are of wooden construction. With stern trawling, the gear is hauled into the vessel over a large ramp through an opening at the back of the ship. These modern vessels are between 100 to 150 feet in length and are of steel construction.

Pros and Cons of Gear Types

Mobile Gear

When stocks are abundant, dragging is a cost effective method of catching fish, especially when coupled with the latest detection technology. On a two week trip, one large trawler with a crew of fifteen is capable of catching 300,000 to 400,000 pounds of fish. Since their gear comes ashore with them, draggers can be directed by the fishing company to exploit a particular stock or species of fish. Draggers have also proven to be the only really effective means of catching significant volumes of flat fish.

On the negative side, dragging has the greatest impact on the environment. Although the proponents of mobile gear technology argue that habitat destruction is not a serious concern, no one maintains that dragging is good for the seabed.

Beaton noted that the landed quality of fish caught on the first few days of a fourteen day trip is questionable at best although fish handling practices of draggers are improving somewhat. He noted that smaller draggers on shorter trips or even day trips have less questionable quality considerations.

The fact that fish companies own almost all offshore draggers and a great many mid-shore boats inevitably limits competition at the port market and disrupts free market forces in establishing fish prices. This often works to the disadvantage of those who own and operate smaller boats.

Another negative impact of draggers, large or small, is the relative lack of selectivity in the size and type of fish caught. Whatever is in front of the dragger's net will go into it. The current state of the technology is unable to prevent the incidental capture of non targeted species, so that draggers often haul in skates, crabs, eels, marine plant growth, rocks and things of that nature. Since the net can't discriminate, regulations allow for a certain percentage of catch other than the targeted species. For example, if a dragger has a ten per cent bycatch and brings in 100,000 pounds of flounder, up to 10,000 pound of the total catch may be cod.

Many efforts have been made to design dragging nets that will allow small fish to escape. At the present time, 5 inch mesh size is the industry standard. In diamond shaped meshes, the holes in the net form a diamond shape that is approximately 2½ inches on a side. Square shaped meshes are also in use, but the general experience of most fishers is that this configuration of net captures one to four pounds of undersize fish for every pound retained.

To build a dragger and equip it with the latest communication devices and detection technology requires a big outlay of capital. Another consideration is the fact that draggers are expensive to maintain and require a lot of fuel.

Western countries have often been quick to adopt methods of fishing from other countries. In this respect it is interesting to note what has happened in Denmark. In a nation where the inventor of the modified dragging technique, Danish Seining, was knighted for his contribution to modernizing the fishery, mobile gear is now prohibited.

Seining is an effective way of capturing schools of fish. The fishers have to haul them in to see what they've got. Undersized fish are smothered in the net and have to be thrown overboard. This is not good for conservation.

FIXED GEAR

Longlining produces a quality of fish that is superior to that caught by any other method. The fish are not crushed in nets nor deformed in any way. They are taken aboard alive and are usually cleaned, bled and iced immediately. The gear is seldom left unattended but, on the rare occasions when this happens, the fish will either remain alive until the gear is recovered or they will escape. In short, longline gear has no harmful effects on the environment.

Longlining is almost completely selective to the targeted species except for some catch of skates and dogfish. Both of these have some limited market potential. Longlining is very fuel efficient, not very capital intensive and quite low tech.

On the negative side, longlining is very labor intensive, especially in the manual baiting of thousands of hooks per vessel. This method of fishing is not particularly size selective, the bait probably attracts some fish too small to take the hooks.

Gillnetting is size selective. A six inch net will not catch any juvenile fish and few if any really large fish. Gillnetting is rather low tech, requires low to modest capital investment and is rather low in labor required. On the negative side, the quality of the fish caught can be questionable if the nets are used in warm water. The quality also suffers if the nets are not tended daily, something which is unavoidable from time to time due to the weather. Gillnets are less species selective than longline and will catch non-targeted species if they are present and large enough to be retained by the net.

Gillnets, if lost, can continue to fish. This is called "ghost fishing." Since the gillnets are constructed principally of monofilament webbing, they could go on fishing indefinitely. The fishers in Petty Harbour, Newfoundland, regard this as an environmental problem of monumental proportions and have banned gillnet fishing in their area. The choice many not be limited to either monofilament nets or a return to cotton nets. A recent suggestion is that the line that joins the net to the topline be made of biodegradable twine. Once this twine disintegrates, a lost net would no longer be able to rise up from the bottom to ghost fish, but would stay on the bottom. Eventually it would be covered with sediment and plant growth.

CHAPTER
2
PETTY HARBOUR, NEWFOUNDLAND

Chapter 2
Petty Harbour, Newfoundland

Lil Clark

It's the uncertainty of it all that bugs me. You don't know from one minute to the next. Where are we going to be when NCARP is ended in May, and the fishery may not reopen for another ten years? Is the money going to be kept going? You know yourself the well has got to go dry.

The Canadian government hasn't got the money to keep up with all of this.

The only good thing that came out of it as far as I'm concerned is that we had time to spend with children, time that we never had when the children came out of school at the 20th of June. That's when we would start off the busiest. And they were put off then with baby sitters, and we were never able to take them out of the harbour. We used to take one weekend off, probably the latter part of August, and take them to the parks or all around the bay somewheres.

I worked in the plant for twenty years when, a couple of years ago, I gave it up and went fishing. About six or seven women here go fishing. We had long days at the plant, 15 or 16 hours at a stretch, when it was busy. When the capelin hit at one time, you'd have to do it. That was it.

It bugs me to hear a lot of comments from other people. They think everyone is getting $406 a week. They think everyone has got it made. Out of 20,000, I think there's only 4 or 5,000 getting the top amount.

I had to take a course in navigation for two weeks. Now I didn't mind doing it. I actually enjoyed it. It's something I'll never use in my life. I can see people given the option of going back to school, but not forced to go. If people want to get upgrading or go take a course, that's a godsend for anyone who wants it. The colleges are so filled up with fisher people that the young people got to wait. I can go out today and tell them I wants to go in September and I'm in.

The only thing stopped by the government is the inshore. The little bit that we take out of the water on the inshore did not cause this tragedy. There's nothing else being stopped. The draggers and longliners that are out there got a ten per cent bycatch. So if they go out and bring in one hundred thousand pounds of flounder or turbot, ten per cent of that is cod. If you have ten boats out, that's one hundred thousand pounds of cod a week. There's no seal hunt going on. The foreign draggers are still out there. I mean, what do they expect? There's still nothing being done as far as I'm concerned. There's a lot more got to be done then take twenty thousand people and give them a bit of money a week and tell them not to fish.

The young people in our community never had so much idle time. That's a worry.

The seal cull should never have been stopped. It has no bearing on the people that tried to stop it. The guy that rammed the ship - Watson. Now I support what he done out there. But he should never have had to do it. Our government should have been out there banging them out of there, not him. But I'm glad someone done it.

They got to put the seal hunt back on. The government is after trading the fish off for every old thing that can come in here.

Doug Howlett

The government didn't listen over the last ten years to small boat fixed gear fishermen. You have to give the fish a chance to spawn. National Sea and Fisheries Products International did otter trawls from January to April when the fish would spawn and they took large quantities of fish.

In the years prior to 1986, before the observers were placed on board vessels, a lot of fish were dumped overboard. They wanted the larger fish for better fillets in the marketplace.

We're now faced with poor year classes from now till the end of the century. You need so much big fish to make sure the stock continues. Selective fishing to catch only big fish is not necessarily a good reason.

LUKE BIDGOOD

They might get it back, but it's going to take a long while. They got to bring in pretty strict rules - offshore, inshore. It's not all offshore caused it. Inshore didn't do their part either, I suppose.

There's always been a seal hunt before Greenpeace stepped in on it this last eight or nine years. Every year they went off here bringing in 100,000. Hundreds of thousand, yes boy. They kept control of the seals, boy. They're numerous now.

SAM HOWLETT

The stocks aren't coming back here for a while. I used to fish handline, not trawling or trapping or nothing. I goes out a scattered time to try and get a meal. You can't get a meal.

Mike Kiely

So, you're an artist eh? Well we're all artists down here - drawing the unemployment. I was at the fishing, but she's gone for good too. What was out there last year, it's gone down 25% lower this year and there was no fishing last year. The seals are looking after the rest of it. The only way it will come back now is if the seals are gone. Greenpeace says seals - they don't eat cod. The CBC was interviewing Peckford. Peckford was on. Peckford's kind of witty and he said, "What do you think they eat? Turnips?"

There's less here this year than there were last year. You could get one out there to eat last year. But now this year you can't get one to eat. And they wasn't at it last year. The fishing was closed last year.

Fishing was on the go here 400 years. They never had no trouble until the draggers came in. They destroyed the grounds. They dragged over the grounds and destroyed everything on the grounds. Spawn or anything like that there.

My funny feeling is to get this back you'll have to stop fishing for 100 years. If there's no women here, you're not going to have no youngsters. As far as I'm concerned it's gone completely.

If the thing goes down to 20,000 metric tons, the seals will look after that. There's about five million seals out there. A seal can eat about 40 pounds a day, is it? If they get hungry enough, they'll eat the gulls.

CYRIL WHITTEN

I was fishing right up to the moratorium. Draggers and gillnets ruined the fishing. Draggers tore up everything and the government introduced the gillnets. The worst about that is the ghost fishing that goes on. The synthetic lines will never rot. The government introduced them. They came so easily. The fellows couldn't get enough of them. If I got twenty, the other fellow would get thirty, the next guy would get forty. The cod nets were of cotton, but for the past twenty-five years all the gillnets were made of monofilament.

If the fish come back I would say that the draggers and gillnets should be banned. When the fish are caught by the gills and are there for a week, they turns red and everything else. The fish would be fresh but spoiled. You see the fish, when they strike those nets, they die almost immediately.

When they put the moratorium on, they should have closed it down more. You go up this shore 100 miles and they are allowed to fish. They thought there was more fish up in them areas. The moratorium only extends to the inshore cod fishermen. If the cod is over for the fishermen around here, that's it. The coast here is too rocky for lobster. Even if they are able to catch the odd meal, what's the good of that? Fifty pound of fillets. That wouldn't mean a thing. With two or three guys in a boat, that wouldn't pay for your fuel. The fishery should be closed down all over.

If people kept on fishing with traps and hand lines, God man, they'd never put a dent in it.

There's lots of capelin. They have a quota on it. Some make their quota pretty fast. In one capelin trap, they got 90,000 pounds in three days. They fish them in smaller mesh traps and with seiners too.

The fellows here are getting a few dollars but they'd be happier fishing.

MIKE HEARN

The dragger fishing has to be completely abandoned. They're raping the grounds, tearing up the bottoms and all the sea plants that's on the bottom, plus the gillnets should be abandoned and never be allowed to put in the water again. The gillnets have destroyed all the mothers - the breeding fish. They gets in places where they can't drag it. Those two measures got to be gone.

We still agree with an offshore fishery, but with longliners, a bait and hook fishery, which is not destructive. You just got hooks running up and down through the water catching fish, rather than a trawler that's dragging over the bottom and tearing up the bottom. We certainly believe in that. That's what's got to happen there.

I can't see fish farming being the real future. The fishery can get back close to where it was if governments allow it. But governments may not allow it. It'll get back so far and bang it's gone again.

Dragger technology has to go. A farmer needs technology on his farm to take up his potatoes or whatever he's got to do but if, in the process of doing that, he's spoiling his ground, that's not good technology. Technology is only good where it's not destructive. The dragger is a destructive technology, and so are the gillnets because the gillnets catch only breeding fish that are moping around on the bottom. And there are so many of them lost, it's unbelievable. There are thousands of them lost. Thousands and thousands. There are thousands of them out here ghost fishing right now.

There's been talk about banning the monofilament lines. Synthetic or no, the gillnets is a destructive measure. It catches breeding fish. The synthetic nets catch fish, they sink to the bottom, the fish rot out and up they go again. Constantly. You can't see them, but you haul them up. Even closer on the fishing grounds.

The draggers usually get their quotas in January and February when fish are congregating together for spawning. They drag right through that. Yet we have people telling us up to today that, dragging through that, there is no evidence that it hurts the fish. The fish are spawning there, and they run through them. It is destroying them left, right and centre. Every animal has a time for spawning and breeding. But the codfish, you can do what you like with it. It don't make sense. You can do what you like with the codfish and it don't seem to hurt them. But it has hurt them. It's hurt them so that now there's hardly any there.

There should be no bycatches right now. Dragger fishing should go altogether. Right now they're out there now. Suppose they're dragging for flounder and he's got ten per cent bycatch. Let's say that amounts to him 100,000, 200,000 pound whatever. He can get that bycatch. If he don't get it among his flounder, he's going to get it anyway. And they say they got observers on the boat. None of this is really working.

The government knows what's going on, but nobody's putting the finger on anybody. They know better than I do, and I'm fishing over thirty years. It's the politics of it all. We have an endangered species. Yet the draggers are still out there. That's because the big companies like Fishery Products International (FPI) and National Sea (NATSEA) and the government are calling the shots. It doesn't matter which government is in the federal government, whether it's NDP, PC or Liberal. Neither will make a difference.

Greed and tradeoffs is what it's about. Right now the government has given out quotas to catch squid inside of the 200 mile limit. Now there's no squid come to our shores the last ten years or more. No amount of squids came to our shore. If they are offshore here taking that squid, how is it going to migrate in? If there's 100 tons there, and fifty taken, there's only fifty left to scatter and come in, right? The Russians have a license to come under the 200 mile limit to catch squid. So how are they supposed to reach shore for us to catch?

Nobody knows the details of the deals Canada makes with other countries. But everybody knows it's going on. We know when we're getting shafted, right? The government runs the fishery like the Mafia runs things. That's what it's like. No difference. Not one bit. Politicians are in the middle of things. They know what's going on. Political campaigns cost a lot of money. That's the way it's all going.

We don't have fish for nothing. If squid became available here now, we'd be allowed to catch it. We might make a dollar. But now with everything put offshore, that's what happened to the squid fishery. The migration patterns don't work. Because you're taking it out here, that squid out there won't make it to shore. If somebody interferes with that, and takes a good part of it, the migration pattern is going to be interfered with, isn't it?

As for the scientists, in some cases they know that it's up and up. And than again, sometimes when they're there, they are under a lot of pressure. In some cases, if they do it independently, they might come up and say the way it is. Sometimes they'll say so much, but they won't say the other thing, because they're being political again and after their own behalf, I guess. It's endless the chain of events that have taken place.

The most plentiful thing we have now is seals. Every species of fish is close to being wiped out. Just before they stopped the seal hunt, we had a controlled hunt. They allowed 60,000 seal pups for the landsmen and 60,000 for the offshore ones. The control on the seal hunt was still going. In fact, a few years before that when there wasn't control, some ships brought in 5,000 and others brought in 200,000. Once, as many as over a million seals was landed here in St. John's. It never wiped them out. If you wanted to cull the seals, you could get away without killing the younger ones, the white coats or first year seals. You could leave them all. You could take second year seals and thin out the seals. In all the years I was at it, it was a controlled hunt. You had a quota and that was it. They knew there was x number of seals out there because of aerial surveillance and everything. In the past years, it was the very odd year in my time when you might see a scattered seal. The last few years, seals were very commonly seen around. You had to go north of here about 100 miles at one time before you could spot one.

The Canadian government is not going to do anything. They're going to these conferences and everything, and nothing happens. Those other countries don't care. That 22,000 metric tons that's outside the 200 mile limit. If they know that's the remaining part of the northern cod stock, they'll just take that. They don't care. Those foreign boats, they don't care. And I seen a piece in the paper a few days ago where it said the surveillance of planes and everything . . . saying OK, there's been no activity inside the line and they haven't been doing this, they haven't been doing that. They got no control of what they're doing out there. Planes are only flying in the daytime, and they only got one or two boats out there. Nah, there's no control at all.

You go anywhere along the Avalon Peninsula where most of the people are familiar with Petty Harbour, right? Now a lot of our communities do well with fishing. They will tell you around Petty Harbour now, you have a job to get ye a meal of fish. They'll hardly believe you. One of the best spots on the island, the most easterly point is Cape Spear just down here. You go out our bay and it's only two and a half miles from us. And you can't get a fish. And this is as prime a time as you can get, right? Well we said two years ago when the fishery fails, when Petty Harbour fails, the fishery will be failed. Petty Harbour has very seldom failed. We got a place for the fish in '91, when I was on the bank. The cod shut her down. We just escaped the bullet from failing, right? They always say, if you can't get a fish in Petty Harbour, you won't get very many anywhere. Unbelievable, really. It's a sad, sad thing.

REG BEST

When we got the 200 mile limit, instead of the government cutting back on the fishing, they gave foreigners the license to fish. The Canadian fleet was also overfishing. There was no one aboard to check on the fish. Except in a few cases, it was just complete chaos out there.

Newfoundland can only depend on the fishery. If it keeps on going like it's going, there will be no harm to call it a rock in another few years, because there won't be anyone living here. There won't be anything here for them to do.

When we were kids, around here any night in July or the last couple of weeks in June, you could go anywhere in this community in the harbour, and you could dip capelin off anywhere any night at all. You'll never do it now. The capelin fishery is going to ruin the whole fishery, the wild life and the whales. They're not going to have any bait to feed on, and nothing to feed their young with. Eventually it'll wipe out an awful lot of wildlife.

To manage the fish properly, all they have to do is use some fishermen to tell them what's going on in the fishery and let them help out. Use the people who know about it, not the fellows who sit behind the desk doing up their little formulas of what should be and what shouldn't be. They definitely have got to cut down on the offshore. New technologies allow the offshore to harvest as much fish as the in-shore with 90 per cent less employment. The offshore effort is going to completely wipe out everything. The majority of us here are trying to make a change, and that's because we see that's our only way of life in this community and in Newfoundland as a whole. That's the only way we can survive.

CORA BEST

I've been in the plant 18 to 20 years. Perhaps I was young and foolish to start working after grade ten. I got my grade twelve in the upgrading.

The closure of the plant has affected me because I don't know where my future is anymore. I'm thinking about retraining in other areas. I find it hard to make up my mind what to do because the jobs are just not there. I'd like to become a nursing assistant because I like working with people and helping people. From what I hear the jobs are not great in that area either. Going back to school was a big adjustment, being a parent. I'm forty now. We have two children. I had to make time for study which I didn't have to before. My husband and children were really supportive. At my age now, I wouldn't want to go into a four or five year course. I probably could do it, but where do I stand when the course is over. Would there be any work or jobs for me out there, right?

The big people who aren't involved in the fishery don't understand how it's affected our life. You get $225 a week and the fisherman is getting $406 in the compensation or the package. We need money to survive but what has it done to my life and my husband's life? It's completely turned it around. We've had a way of life for so many years. My husband's been at it all his life and I've been at it all my life. And here we are now - we don't know where we stand.

They just closed the fishery down and gave us that, as far as I'm concerned, just to quiet us. We don't know if there's ever going to be a fishery again. My husband has been saying for years that something should have been done about the economy and fishing spots, but they waited too late.

People seem to forget what's happened to people's lives. Just being under the stress of saying, "What do I do? Where do I go? Is there anything out there for me now that there's no fishery?" I'm hoping and praying that the fishery will come back. Probably not for my children because I wouldn't encourage them to get involved in the fisheries. I would encourage them to go and get an education. But for the people that are left and depending on the fishery, I'd like to see it come back for their fulfillment and their needs of making a living. It's a way of life. It's been there for so many years. My father was a fisherman. He's 73 years old right now. He's retired, but still, like he said to me before, "You know, I'm so used to getting up in the morning and seeing the boats come in with the fish, and going to the plant and selling it and the day sort of fulfills itself." They went fishing and got the fish for the day. All that's gone now. Even just to watch the boats come in with the fish was a joy to a lot of older people in Newfoundland. With no fishing industry, there is quite a blow to the fisher people. I talk to my friends and my family, and people are saying on these open line shows, and people are calling in and they're criticizing fishermen and plant workers. They just don't understand what the people do. We got to have money to survive. But we want jobs. I'd rather go to work in a fish plant and earn my money than go on the grants. If I was to say, well you take that now, and we'll decide in the future where you'll all go and what happened to you, right? There are a few bad apples in every barrel, but don't class us all the same. I mean my husband really enjoyed fishing. He enjoyed on the water and it's in his blood. He grew up that way. It's not that he can't do anything else. He can get by with lots of different jobs, but he likes fishing, and that's what makes him happy. And if you're happy on the job, as far as I'm concerned, that's what matters the most. But like regardless of going back to school and everything, I feel good about that too. Probably if I get out now and retrain in another area and get a job, it's probably the best thing that ever happened to me. But I would still like to see the fishery return to the people that really need it. A lot of things have been turned around. They're forgetting the social impact that it's having on the fisher people. The main point, the person that was involved in the fishery, is being overlooked. You get people saying, "Take your money, and if you don't want to do anything with it . . ." That's totally wrong. People don't want that. Newfoundlanders are not known to be lazy or just sit back and take that money and not do anything. Another part of going to school is what I found has made me able to cope with this moratorium on the fisheries. Going back to school has given me something to do rather than sitting home in the daytime without anything to do. You know, like I found that has helped me a lot. If I just had to sit here all day long and just not do anything, I think that would make me feel useless. But I can be with friends and meet with different people. It's given me something to do. But now in the meantime, like I say, I've enjoyed going back to school, and I'm proud that I have my grade twelve diploma now that I didn't have a few years ago. It's taken a while, but at least now I've got it. And hopefully, I'll go on and retrain somewhere else. But the fishery will always be part of my life no matter what I do or where I go. That's just the way life is around here wherever the fishery goes on.

I guess I'm angry at the people who are responsible for letting this go on. I was tired of hearing my husband and other fishermen say: "When are they going to wake up and do something about it?" They were saying that for the last ten years, that something should have been done. Nothing was done about it. Here in Petty Harbour alone, like I have a son - from the time he turned 12 or 13 like in the summertime he'd be cutting out tongues down around the wharf and probably get a job when he grew up, in the fish plant. He could go out in the boat with his dad and get a few extra dollars for going back to school, and especially like now where he needs more money than he ever needed, regardless if he's going to university. That's all been taken away from him. The last couple of years that he spent in the plant, I think that's what he was doing - saving the money for his university. And that's all been taken away. As it's turned out now, he's been fortunate. He got a job on a recreational plan this year and that's good in a way, but I don't know where he stands next year or the year after. He's going to need that money for school. My daughter is not going to get any work there at all. She's thirteen now, and with the fisheries closed down until next year, she's going to miss all that, and she's not going to be involved in it at all. She's not going to know how the youngsters used to get their money a few years ago. My daughter won't experience in the fishery what Tom did, and what I did and what her father did. And I feel that growing up as a Newfoundlander, she'll know, but she won't have the experience in it. I encourage her to go to school and, please God, do something outside the fishery, but I'd still like her to be able to experience part of the fishery.

To just stay home and think about the problems, what happens to me if they cut out this money, what do I do with bills to pay, mortgages on the home, whatever. It's too much to stay home all day long and dwell on all these things. That's why I think the retraining is good. I'm saying that I'm hoping that it will be good for me, that I will retrain in a certain area and get a job and it'll be a benefit to me that way, but like the people that are just going to stay home and not do anything, I pity those people. It's going to have a tremendous effect on our lives. It's only a year since this has happened. Last year they didn't know if they were going to go fishing or not. They were getting their boats ready to put them in the water. Not this year. They knew they were putting their boats in the water to protect them from rotting and everything, but they knew they weren't going fishing. It was like a dream. You wake up and think, "I'm going fishing today." Then the reality strikes you that you're not going.

About two months ago my husband came home one evening and said, "I'm going down on the wharf now." And he came out with his plaid shirt on and his jeans. I said, "My, Reg, you look like you're going fishing." And he said, "Cora, I wish I was." When it's in your blood, it's hard to accept the fact that you can't go.

Hopefully, it will build up and come back. It's going to take a while. The people in their forties and fifties might not get a chance at it.

Reg's uncle, Arnold Chafe, was ready to retire but he didn't get a chance to retire on his own. He's retired now. Take your package or whatever. I think that's sad.

TOM BEST

A lot of our trouble was caused by the technologies that were being introduced and promoted. We still don't see any great changes in the way governments are handling this. Basically, there's a lot of games being played, a lot of smoke and mirrors and things like that. I think our govern-

ment is saying now that we are being successful through the diplomatic approach in moving the foreigners away from our coastline. The hard facts about that are the only reason foreigners are moving away is because there's no fish here for them and it's not in their economic interests to come over here anymore. The numbers are still in the seventies and eighties of vessels fishing out on the nose and tail of the Banks illegally. And then we have all kinds of joint venture agreements with Canadian companies inside the 200 mile limit. But while they're classified under utilized species, most fishermen in eastern Canada, particularly Newfoundland, would argue that there are no under utilized species. And some of the ones that they have identified as under utilized are just about wiped out as well as the cod stocks. Even turbot and things of that nature down on the Labrador coast. But they are not listening. They have still identified it as a big quota pool of under utilized species. Newfoundland and Nova Scotian companies and others are participating, through arrangements with the federal government, in joint ventures with foreigners to catch these species. And under the arrangements and conditions, they only have to land ten per cent of their catch, contrary to what you may hear about landing it in native plants. The requirement in the under utilized development pool is that you only have to land ten per cent of the catches that you catch. What the ventures can do with the rest is that it can be sold on board to the foreign partner and all this kind of stuff, and it's still going on.

So there's a certain amount of trading off for other concessions. The government tries to present itself as not doing that. It's all about trade and other trade considerations that are more in the interest of sometimes political friends, I suppose. There's a lot of big corporations in central Canada and Quebec and places like that. That's what it's all about. Now there's a number of us in the industry who know that because we've taken a lead in these matters and learned a lot about what they are all about in the last eight or ten years, but most fisher people are not involved in these matters. They sit back. Complacency has set in. Dependency and all that sort of stuff. Right now, because of this program we have, the moratorium and the support program, the compensation package that's there with it, it's the hardest and the most scary time ever because most people are not prepared to stand up and take a position and do anything unless it's something that's affecting them right here and now. For example, if they were trying to sell a boatload of fish and had no place to sell it, you'd get a hundred people in a meeting hall. But as you know, right now there's no fishery going on and they're receiving cheques from Ottawa - you can't get people out. When May 15, 1994[1], rolls around and if the moratorium gets extended and with no support program, you'll get them out in the thousands again to the meetings. So this attitude has all set in, this dependency and sort of a realization that we don't have much of a future here but, due to dependency, we're not prepared to do anything about it. It's bad as far as I'm concerned because I haven't been one to sit back and not take positions.

The general public is not in touch with the issues. My wife is working in the education system coordinating one of these ABE (Adult Basic Education) programs, and we've had 70 something people enrolled. We only have 130 fishermen in this community. That's a lot of fishermen for one particular community. But we have close to 300 eligible Northern Cod Adjustment and Recovery Program (NCARP) recipients as they call them, fishermen and plant workers, who are receiving benefits under the program. Now because of our organized approach over the years, and our unique structures and systems in place through the co-ops, (that's not to suggest that every member is a hard-line co-op member) we've been able to get 75 to 80% enrollment in this community in NCARP sponsored training education programs. Across the province, they've only been able to manage overall 10 to 15%. Most communities don't have any organizations, don't have any systems. Fisher people may be members of unions, but then they don't have systems in place in their community, organized structures and things of that nature. They may have a business rep, most of whom with the unions around here have been chosen from a small group rather than elected by the majority of fishers. So they don't represent for the most part the interests of the majority on many occasions. So we fishers have a bad problem because most communities in this province are distant from St. John's, where all the decisions as it pertains to how things are going to happen in this province, for the most part, are made. And then everything else happens in Ottawa when it comes to the fishery or is somewhat connected to the federal government system.

The fishermen here tried to promote their own inshore local a few years ago when the unions split down the middle. The United Food and Commercial Workers (UFCW) local, that was the fishermen's union here, at that time became part of the Canadian Auto Workers. And many inshore fishermen rallied to try and get their own inshore local because the union wasn't taking positions on matters that were in the long term interest of the resource. The reason for this was because they represented both the inshore, and also the offshore dragger people. We've had a major, major battle with trying to get them to take positions on quota reductions and things of that nature because of the fact that the offshore sector was more year round and provided them with more consistent revenues. We tried to get our own local with the UFCW when they offered the opportunity and it didn't work out. The other side had been around more than twenty five or thirty years and was better organized. We just missed by 1000 votes or so. And so there's all kinds of problems with the unions. We blame the fishermen's unions in this province as much as anyone else for having us in the state we are in because they didn't take the positions that they should have and could have taken. And they are still not doing it today. When Paul Watson came in here, even though he created some major problems for the sealing industry, he went out and did something to focus world and local attention to the problems we have. They tried and were successful in discrediting him. We have to get the environmentalists and everyone else on side. Many of us supported his effort, but then our support was token support because of the concerns we have about his sincerity. But we saw an opportunity to discuss with him regarding increasing seal herds and the impact on the eco system and all that sort of stuff. The fishermen of this community held a public rally just to get some media attention to show our support for Paul Watson saying that the government is not doing enough. This person is doing something now, so this is what you should be doing. Our stand upset some people, and they went right against him. We have some complex situations around here when trying to deal with critical issues.

So there are some hard, cold facts and other realities that haven't set in yet among the people. And what we're doing in our association, the association which I am president of, the Association of Newfoundland and Labrador Fisheries Co-operatives, is to present the realities to fisher people and encourage them to take some action. The co-ops are a bit more unique than anybody else in the fishing industry of this province, because fisher people for the most part own those operations, and they use them as a service, but basically they still work arrangements where prices are negotiated the same as they would be by unions or anyone else. They sell to their own operations, but they are member-owners and things of that nature. They are a bit more unique in that their membership understands all aspects of the fishing industry where most others in the fishing industry don't. Plant workers have never, ever been concerned about the resource. It's only now when they start to lose their jobs that they realize that something has happened to the resource. When we needed plant workers over the last 10 or 15 years to get them on side, for the most part we were arguing against plant workers who were looking for bigger and bigger quotas especially in the offshore towns to keep their facilities going. In the midst of all the problems, we still have inshore plant operators who are participating in foreign, joint ventures, while pointing the finger and blaming all the problems of the industry on the offshore companies. A lot of hypocrisy goes on in the industry. For the most part, a lot of the people out in the communities just function as individuals, either fisherman or plant worker. In the case of the fisher- men, they are all individual owner-operators, really. They depend on those fish plant operations as a place to sell their fish, and most of the plant workers depend on those operations for jobs. And we've been provided, both the fishermen and the processing plants, with the technology . . . and I'm including myself because I'm a fisherman and have been for thirty years . . . with the technology, or the tools you might want to call it, to destroy ourselves. That's what has been done because all of them are individuals and competitors, when we talk about the harvesting end of things. At the other end of the problem, we have so many plant workers who don't know anything about the resource and how critical it is to their job. And it's only when they lose their job that it starts to sink in. So it's been a hard battle in trying to get the message through about the things that are happening around here, things that are critical to their very survival down the road. Not everybody in other sectors outside the fishing industry can relate to the importance of the fishing industry to their jobs. There is a small percentage who understand, in particular the business owner himself, but their employees don't.

This province was settled for one reason and one reason only - because of fish. Everything else sprung up around it. The commercial centres sprung up to support the fishing industry, but the people who are employed in those commercial centres can't relate to that for some reason or other. And I don't think they ever will, even though the government had to cut back so much here the last couple of years. It had nothing to do with all the fiscal concerns due to global economic problems and things of that nature. The fact is that for the last 10 or 12 years in this province, plant after plant after plant in the fishing industry has been having problems, and many are closing down because of limited resources. Then people are transferred over to provincial government support programs, welfare and whatever else, if they were lucky enough to be eligible. They've gotten to know how to work the system where they now get ten weeks unemployment insurance contributions and revert over on federal programs. But an awful lot of people in the industry don't have a lot of money right now. And when they don't have the money to spend, obviously people lose their jobs in other sectors of the economy. But people in other sectors can't relate it to problems in the fishing industry. And they're failing to relate to it again. Right now we can't get the message through to the public pertaining to how serious the situation really is, even to the educators as unbelievable as it is.

One golden opportunity has sprung up as a result of this moratorium because a lot of the educators are involved in delivering adult education programs to fisher people, and now they are beginning to realize, as we did, the importance of the fisheries. We were the first to identify adult education programs as critical to the whole process of NCARP training and education opportunities. From our perspective, all NCARP recipients should be looking at training opportunities. Government is now doing another information program, the intent being to go back out and try to present fisher people again with the realities of why they should really seriously consider enrolling in some kind of education program. And now they're putting the focus on adult basic education more so than anything. While they are doing that, this two-week module has been developed and added. It's about fisheries and how it fits into the provincial scheme of things and the Canadian east coast scheme of things as an economic support base. You wouldn't believe how far out of touch they were in designing the response program because we were the people that identified that module as critical, not them. The educators didn't identify that as critical. Now it's starting to sink in. I've been at some of the sessions where they've had these workshops, and it's like they are flabbergasted to hear the things they're hearing about the fishing industry, and how the wheels turn and how it impacts everything else. So they are being educated. They are going to go out and try to do a briefing process, a kind of more or less motivational process, to get people involved. So that's happening now as a result of the lack of interest in the training and education component of the program. And a lot more educators are starting to say that we're going to have to get things into the school system about the fisheries and where the fishery is positioned in this province, and how important it is to the economy. But individuals in the fishing industry have related to that because they have had to get involved in the issues over the last ten to fifteen years, and they've certainly learned how the whole system works.

Most people in the fishing industry do not have very high levels of education because, in communities like this, at 14 or 15 years of age, they started to work in the fishery. Even myself, I quit school about a month before I finished grade eleven and then went back in and took a supplementary exam and graduated. I also learned another trade while I was fishing. Not too many people did that. A lot of people got out when they were in grades 8, 9 and 10, got in their boats because there was good money to be made and stayed there. So you can imagine what academic level they're functioning at now - a lot of them ten, fifteen, twenty five years later. What we have found is that very few people in the fishing industry have high levels of education. With more education, these people can make better determinations about their future, whether it is in the fishing industry, if we have one in the future, or whether it's somewhere else. But if they don't do that, they have no alternative because, first and foremost, if there are some trades that have some potential, some course offerings in trade areas that may have some potential somewhere, they're not eligible to enroll because they don't have the levels of education to qualify for enrolment. So these are the messages that we're trying to get through. Some of the agencies involved in this process were only out selling courses for financial benefits. That's the way it all started out. It was criminal as far as I was concerned. And we've made some strong statements about that matter right from day one.

What we've seen around here in the adult education programs is that, when people get in, they become a lot more self-confident. Fifty years of age today is not that old as far as I'm concerned, and individuals should take advantage of various opportunities. What we're finding is an awful lot of interest in computer courses. If you go back to find work in the fishing industry, every plant will be more and more mechanized. All the boats will have far more sophisticated equipment. There may be a legitimate core of fisher people identified. Unions are using these arguments in a strategy to sell courses because they've got a lot of government contracts. Some of these courses are meaningless to an awful lot of people. And some of it is tied to an argument towards professionalization for fisher people in the future. We're saying and have been saying that we should identify a legitimate core of fisher people. And those who are not legitimate shouldn't have access to the fishery like they had in the past because there has to come an end to when you can jump out of any trade or any job and into the fishery. As I mentioned, I have an electrical trade and I would never go out competing with other people while fishing because I know what that means to people. I was learning that trade as I was fishing. But as I was learning that trade during my teenage years, I wanted to continue fishing. So I wasn't interested in the electrical trade even though there was lots of work on the go at that time. I was doing better at what I wanted to do. However we have a lot of people in this province who gear their holidays and annual leave to the fishing industry, and we've had that going on for years. They've done all kinds of damage to the legitimate fisher people, when it comes to selling and prices and all these sorts of things. I think that some of these people who have other trades are discouraging fisher people from getting involved in education programs. That's the biggest mistake displaced people could ever make. If there's no fishing industry, what can they do? They have to better educate themselves to give themselves some trade opportunity, whether or not they ever use it.

Too many people believe that the fishery is going to come back and will be here tomorrow, but that's not going to be the case this time around. As far as I'm concerned, it's gone that far, and I have never been a pessimist about the industry. I've always kept fighting to try and get better measures implemented for the long term. I would never, ever, given the current situation, encourage anyone coming out of high school or coming though their teenage years to choose fishing as a future opportunity. It's just not there anymore, and it's not going to be there for a long, long while. If the government is going to continue to do what they are doing right now, and they still haven't changed . . . and I've just seen a document that the provincial government prepared called *Changing Tides*. I have a copy of it in my briefcase. Their strategy on the fishery of the future is exactly what their strategy has been for the last 30 years, and that is a downsized fishery, more centralized, more efficient, but efficient from the point of view of corporate mechanized operations in the fishing industry. They say they will strongly support and give priority to the inshore fishery as the stocks rebuild and as access to those fish becomes available again, but what they're saying in this document in effect is that they are going to promote the kind of technologies that will allow people to go further afield. That's the very thing that's got us in the mess we're in right now, the various types of mobile vessels using destructive technologies.

The proper balance has never been there, and the management of the technology has never been there. I've been arguing across the table with all of the offshore representatives. Everyone of them has been chosen, by the way, to sit on those councils and commissions government have set up to steer the direction of the fishery of the future. People like ourselves didn't get appointed. There was a group called the Atlantic Groundfish Advisory Committee (AGAC) whereby everyone in the industry from the different sectors had a voice for the last ten or fifteen years, and the only loud voice against the direction that governments were taking, and all the lobbyists representing the offshore, was myself and a few others representing the Newfoundland Inshore Fisheries Association (NIFA) or the Fisheries Co-ops in this province. We were taking hard line positions against the arguments that were being presented for increased quotas, and everyone who sat around that table with the exception of us have now had representatives appointed to what is called the Fisheries Resource Conservation Council (FRCC) and the Foreign Allocations in Canadian Waters Committee (FACWC). They have discussed three new commissions. They were talking about setting up an Atlantic regional board. Two are staffed right now with the senior vice president of Fishery Products International, appointed chairperson of the critical one, the Fisheries Resource Conservation Council. And every individual appointed to the new councils, for the most part, is some high paid executive who sat around the AGAC arguing against our position on management issues. The only people left out are the inshore voices, with the exception of the Fisherman's Union that has been appointed there, and they represent both the inshore and the offshore. The fact is that they've never taken a position, nor did they ever take a position in the last ten years, on resource matters that were critical. My concern is that we are still going in the same direction. The documents that I'm seeing on the strategies for the fishery in the future are all geared towards enhancing technology and mechanization in the inshore, providing increased ability to move further from the coastline in middle distance waters. The arguments that we used to keep the foreigners out in 1977 were the arguments that the technology that was being used in our waters by foreign vessels was destroying our fishery. You would think that someone would be intelligent enough to say, "Well look, if that's what destroyed it, why should we promote and reintroduce the same kind of technology?" But that's what they did. They couldn't wait to get out there. The emphasis wasn't on protecting the resource but to take the fish where the foreigners were taking them; but now to take it with Canadian companies using the same technologies. They missed the boat on what they were doing. I'll tell you, as far as I'm concerned, this was a stupid approach to fisheries management.

Anything that you're tearing through the ocean bottoms is having an impact on the areas where fish probably would have concentrated for various reasons such as feeding and spawning. Here in Petty Harbour we have specific regulated areas of water. Some of the areas are way better for what we call the baited handline fisheries than others with shoals and ledges springing up all over the place. Fishermen using baited handlines anchor in various positions depending on how the tides are running. Boats anchor a certain distance from each other. There's a specific understanding of what we do with respect to anchoring a certain distance from each other when handlining in the same area. Fish concentrate on some of those shoals depending on temperatures, and depending on the plankton that's around, and depending on the availability of bait and other factors. There's no dragger technology allowed in this community, by the way, but this is only one little community in the far bigger scheme of things. And there's no gillnets allowed here. We believe gillnets are even more damaging than the draggers, by the way. That's the cod gillnets, the bottom gillnets. And they are banned here as well, and have been since their introduction. So the draggers are out there in every other area except ours. They wouldn't be able to drag here anyway because there are too many shoals springing up. But they are doing major damage elsewhere and, in addition, destroying the ocean bottom terrain.

They introduced gillnets in the late 1950s to the fishing industry. Again it was a transfer of technology that was being used by other countries. It seems we're always following the direction of other countries and destroying ourselves. They've done the job on themselves, then they move over here in what is referred to as distant waters, and we promote and encourage the same kind of technologies that they used to destroy their fishery in their own countries. Gillnets are of a synthetic monofilament material. First, when they were introduced, they were just old cotton nets that could rot away or get on the ocean bottom for a while and deteriorate. But that hasn't been the case for the past fifteen or more years. I would take a stab and say that millions of them are out there right now ghost fishing. With the exception of this community, everyone used them and they were introduced as a part of fishing practices all over eastern Canada. And as they deteriorate or get torn, or anything like that, fishermen, as careless as they are on many occasions, not looking at the future, either discard them or throw them overboard. This is of particular concern out in middle distance waters with those modern type 50, 60 footers in the last eight or ten years as they get farther afield, using them on the nose and tail of the banks and the virgin rocks and anywhere else where there are shoals in the mid-distance waters between here and the 200 mile limit. Most of them have to run out of storms a lot of the time, with hundreds and hundreds and hundreds of them being used by a single boat, and they don't get time to take them back because of the storms and everything else. When they go back they can't find them. They just busted off the floats and sank down on the bottom. These nets are always on the bottom, possibly millions of them in the ocean right now, fishing away continuously. These nets are anywhere from three to four fathoms, five to six fathoms deep, to thirty fathoms deep. They have corks on the top section, and they have lead ropes on the bottom. The lead ropes stay on the bottom and the floats stay up so when that fish is swimming over the ocean bottom where it is a lot of the time, especially the bigger, mature fish, these nets intercept them. People will argue that these lost nets will sink down after a period of time and they'll tangle up on the bottom, and they'll deteriorate and all that sort of thing. That's only a self-serving argument by people who are using them. What happens is they will fill up with fish, the fish will rot, the net will get heavy and it will sink down to the bottom. After a period of time, all the fish are completely deteriorated with all the bones and everything, and up comes the corks again and the net continues to fish. We've argued and argued and argued to try and get the gillnets banned from the fishing industry in this province. We are the only community that has them banned. The scientists are trying to figure out for the last four or five years what happened to the spawning biomass that was supposed to be there. They estimated two years ago that it was anywhere between 70,000 metric tons to 150,000 metric tons. Now they figure there's only about anywhere from 15,000 to 20,000 metric tons of spawning biomass left in the ocean. You couldn't believe it, but that's just about nothing. And a lot of them are still being fished out by gillnets that have been discarded. In 1989-1990, when the fisheries got so bad in the other eastern provinces and all along the other coastlines of Newfoundland, the only place that was virtually untouched for the most part with respect to gillnets and draggers, is the area off here referred to as 3L. This is a fishing zone. This is where the scientists identified that, as far as they were concerned, the whole northern cod biomass was concentrated from 1988 up to 1990, 1991. Suddenly they started preaching that, and as they preached and the word got out, with no controls on the buildup of the gillnet fleet, we moved from five gillnetters in 1985 to 200 licensed gillnetters fishing this area. With a million dollars worth of technology on many of these vessels for finding fish, everyone of them concentrated here in this 3L area in particular from 1989 up to 1991. We presented arguments that if they didn't control and stop the buildup of this fleet, that the spawning biomass they were suggesting was left in the ocean there three or four years ago was going to be just about completely wiped out. They didn't listen to us. Now they are coming up with all kinds of arguments and reasons as to why they can't understand where the spawning biomass went. In my report that I just did as a brief to Crosbie, I outlined the concern we had over the buildup of that fleet and the fact that we strongly promoted the control of it. Dr. Les Harris had just done a report for the federal government and all he did was document the same thing that I'm telling you right now. The only flaw in his report was that he thought the way to go, for the next number of years, was to only allow the bigger fish to be taken out of the system and allow the small fish that hadn't yet reached the reproduction stage to continue to build up to allow them the opportunity to be part of the spawning biomass. What he forgot to realize, given that he's a bit more distant from the fishery than he thought he was, was that we're still catching those small fish and it didn't make any sense whatsoever to be promoting and suggesting that we could use the kinds of technology that could take the bigger ones out of the system until those small ones got to the stage where they could ever reproduce. The one flaw in his report that I point to is the fact that he promoted selective fishing technology that would take the bigger ones out of the system. And this is why we didn't get the control of the buildup of this gillnet fleet. His report was taken up by the regional director general of Newfoundland and his recommendation on the measures that are in there could either be implemented by the federal government or not. It was supposedly accepted, but very few of his recommendations were implemented. Eric Dunne, the regional director general, took this report and made an additional recommendation that was supposed to control the buildup of the 3L gillnet fleet, amongst other considerations. Because Dr. Harris inadvertently supported gillnets and had so much credibility with government, and because the union had all kinds of problems in supporting the control of the buildup of the gillnet fleet because it affected another sector of their membership, we didn't get those measures. I went to a recent fisheries forum held in St. John's, where scientists indicated they couldn't understand the disappearance of the spawning biomass, and I just got up and said, "Look, don't tell me that we didn't tell you what was going to happen to the spawning biomass." And basically it just gets shoved under the rug and it doesn't become a public issue. The media don't even pick up on it.

Back when they were introduced, gillnets had an 8 inch mesh size. An 8 inch mesh is 8 inches diagonally. If you pull the mesh tight together on a diagonal, it's 8 inches straight across. It's a big net. That size could catch up to a 35 inch fish because they only mesh in the head and that's where the gills get caught. They started with 8 inch gillnets all over the province. In the late '50s and early '60s, and within a period of four to five years, they had to go to a 6 inch mesh because the mesh was too big and the fish that were left were just going through them. All the big ones were being taken out. And then they went to 5½ inch mesh and last year there were fishermen, even given the bad situation with the resource because they were catching less and less fish, arguing to have that mesh reduced to 5 inch mesh. A five inch mesh gillnet can take anything from a 12 to 16 inch fish. And a 16 inch fish is only a three year old fish. And a fish that can reproduce has to be seven years old. It takes about a 20 up to a 25 inch fish to be a reproducer, and we've taken basically all those fish out of the system. Now we have probably millions of those gillnets left in the ocean around the coastline ghost fishing. Then you've got the technology in the middle distance vessels moving further afield using gillnets - even though these vessels were supposed to be developed using Scandinavian longline techniques, longline trawls, etc. The longliners, the 45 to 65 footers, were being built for limited entry fisheries and for a small group of fishermen in our province. They were geared up with gillnets and, as they destroyed one thing, they went to something else. They have the technology, state of the art modern day satellite type tracking detection technology, that will find every blemish in the ocean. Wherever they find schools of fish, if there's any left, they're using gillnets. They can sink them down on the bottom regardless of rough bottom terrain. A dragger can't do that. That's why a gillnet becomes ever more dangerous than a dragger. You can only use draggers in more or less smooth bottomed areas because, if they tear across those ledges springing up from the bottom, they would tear the drag nets to pieces and wouldn't take back anything, and their productivity would be very low. So they can't fish them in certain areas of the ocean, but you can put a gillnet down anywhere regardless of the underwater terrain and things like that. That's why they become most destructive. And the second reason why they become the most destructive is because fishermen, careless as they are, just discard them over and over again. It's cheaper to dump them overboard and go in and have new ones made than to repair them. The government is talking about developing gillnet material that will break down, but they haven't done it yet. So they are out there, and they will be there for hundreds and hundreds of years. I guess with the tides and everything else swinging them back and forth for other reasons, they will eventually break down, but it's going to be an awfully long while before they do deteriorate to the point where they won't fish. The people who use them argue that they won't fish after they lose them on the bottom, but still and all, they fished while they were using them. So there you go.

So we don't have right now, as far as I'm concerned, a lot of hope for the recovery of the stocks. Now you can see the importance of educating the people. There are not a lot of people who involve themselves in the issues. There are only very few. This community is seen as the most united and the most organized fishing community in the province. The fact of the matter is that there's only one person speaking for this community on most occasions, and that's been myself for the most part. Prior to me, there was probably only one person - usually the head of the fishermen's committee. Very few people come along who are prepared to take the lead. I've heard the argument by people that there's always someone who will come along to fill the gap for someone that's moved on, particularly people from Memorial University. That doesn't turn out to be the case, unfortunately. We have a twelve person fishermen's committee here, and a twelve person board of directors of the co-op, a lot of whom are committee members and things of that nature, but when it comes to them committing the time and effort that is necessary to present the messages that have to be presented, believe me, they don't. Talk about the co-op out there, we're struggling now to keep that alive. The co-op is strictly a cod fishing plant. We built it back in the early '80s and, ironic as it is, we had more fish than we could sell because the plants in the area had geared up to process capelin, which is again a very short-sighted, lucrative fishery. The capelin stocks are critical to the survival of the whole eco system, but we're still fishing away at them, and people can't see the danger at a time when the stocks are dwindling, the seal herds are increasing, and all these sorts of things. Well, they see it I guess, but a lot of them don't really care. They make a lot of money today and they don't worry about down the road. Most of the people who get involved with those kinds of arrangements are short-sighted and self-serving. The problem with the banned seal hunt is that it's being used too much as a rationale for the problems we have in the industry by those who are serving themselves. That's dangerous as well. That can become another bunch of smoke and mirrors about the cold, hard facts about why we've got the problems we have right now. Some of these environmentalists, and I hate to agree with them, have a legitimate argument about the fact that the eco system balances itself out over time. As one thing goes up, another thing goes down. When something goes down too low, the others that went up will starve to death, and then all these balances will take place. But when man gets involved, he's disrupting everything again. So with man's interference, when certain things get out of whack that shouldn't be out of whack, you've got big problems. The reason I would support Paul Watson on this particular initiative on the foreigners is to get attention to the issue of overfishing, but also to try and open the door for another argument. He will argue that although the seal herds are increasing, they are not eating cod. I can probably agree. I mean there's been no seal fishery around here since Watson and Greenpeace. I agree that the seal herds have been increasing enormously in the last few years, but he suggests they will die off because if they can't find enough food, nature's normal balance will take place. What he's forgetting is that we still have all the industry, and we can't get them on side with respect to stopping the commercial capelin fishery. So while the cod stocks are at such a low level, they hardly have anything to eat. Seal herds are increasing to the point where they are competing aggressively for what food is left in the system in the food chain. Their major diet is capelin as well. At the same time, while seal herds are increasing dramatically and competing with cod for capelin, we have a major commercial capelin fishery taking place. Most fisher people would argue that the capelin stocks are in bad enough state that they should have been closed down eight or ten years ago because they are critical to everything else in the ocean. Cod needs capelin to feed on. The seal herd is increasing to the point where seals are changing their eating patterns. They had a variety of things to eat, but the biggest part of their diet was capelin. There are so many of them out there now that there's very limited capelin around. They have become a major predator, a savage and aggressive predator, for the bit of capelin that's around. Cod are not going to get access to those capelin while there are seals trying to feed on them. So it's great for Paul Watson to say that seals will starve themselves to death over a period of time, and nature will reduce their numbers. Unfortunately, what will happen is that while they're doing that, given the intrusion of man, the capelin resource will dwindle to the point where there's not enough to go around for cod and seals. What's happening is that the seals are aggressively competing for less and less capelin and cod are not getting a chance; and if cod don't have capelin, cod will just die off. Period. They don't feed on anything else with the exception of a very small percentage of their diet. If you get the door open with people like Paul Watson to start talking about these things, and if they're so sincere about the eco system and the environment and everything else, and the cod stocks, they have to realize that arguments about nature balancing itself is disrupted right now. If you don't allow the seal herd to be culled to allow cod to feed on capelin until they get to the point where they can rebuild themselves to a sustainable level, the biggest problem here is that you're going to destroy the cod stocks in the process, and they'll never get the chance to rebuild so that they can become reproducers again. Now, that's only one part of it.

We have all the other things that are going on in the industry that should never be happening. He needs to understand as well, if he's so sincere about what he's talking about, that even though nature balances itself out, we have still catastrophic levels of capelin fishing going on. So perhaps he should start focusing his attention on not only foreigners on the nose and tail of the Banks, right now for whatever reasons he's doing it, but support our arguments that the commercial capelin fishery should be completely banned altogether. If he wants to use the arguments that seal herds will balance themselves out and everything else, he's going to have to realize as well, if he's so concerned about seals, that they don't have anything to eat because we're still fishing capelin commercially. If he doesn't take a position, then he's doing to seals what he's arguing that we are doing to seals. He's allowing the killing off, only in a different way. They're not hitting them with a club but rather taking away capelin which is the seals' main diet. But if we can get the environmentalists on side, they can help change the direction that man is going on all of these issues because they can get public attention quicker than most. While we have all these problems, all kinds of activities are taking place in our commercial fishery just as carelessly as in the past, while pretending to the general public that there is concern about the state of our North Atlantic fishery.

Nobody in the media is writing about all these things and there is no process of disclosure, with the exception of things like I'm doing, writing briefs. I honestly don't believe there are a lot of other people in the industry who are doing the same things. I take the time to jot down my concerns and positions on matters and, before presenting them wherever, I consult with a group of fishermen to see if they agree. I never present any position on behalf of fishers unless they agree. But when presented to very restricted audiences like the Fisheries Resource Conservation Council, they may decide what they want to do with it, modify it or something else.

The Fishermen's Union is in a difficult position. They are supporting themselves now from government contracts and training and education for NCARP recipients. Because of this, they are very restricted in what they will or won't say about what should or shouldn't be.

The people in Employment Immigration Canada have now seen how people are taking advantage and manipulating. There's $300 million dollars cut out that's a part of this program for the purpose of NCARP training and education initiatives. There may be $100 million used yet to this point because the facts are they can't get a big enrolment. They thought it was going to be easy, but it wasn't. We told them certain directions they should follow, that is in the fisheries co-op sector, but they didn't listen, and May 15, 1994, is quickly rolling around. The same type of program may not be in place when this particular one has run out. But they're finding that it has become a big scam in the case of a lot of training agencies and unions. Some of the things that the union has been promoting have not been in the interest of individuals, but it has been in the interest of sustaining the union and the executive in particular. They've taken a rap for that over the last two or three months. It's all been kind of kept quiet behind closed doors. They're changing their direction and their strategy, but it's only because they've been chastized. But they still got the contracts. And all they are doing is changing their approach. Some of the things that they were promoting have been by now just about scratched. Some of the things they were promoting towards professionalization. Anything they could get out running as a course, they ran it just to try to get people in. They were charging for the courses and the enrolment and things like that as a part of their contract. But the more courses they could sell, and the more course outlines and whatever else - different options - the more money they made. That's the way the contracts are set up. If you don't get enrolment, you don't get money. Fortunately, there are some good people out there in Canada Employment who have identified this stuff, and they're in key positions, but still it's all being kept quiet. There's an awful lot of abuse going on with this particular program. Not so much on the part of the people who have the opportunity to participate, but on the part of the people who go out and deliver training and education for profit. We've kind of stepped in the middle of that and made some public statements about it. Either you do this right or get out of it. Don't go out there using people. Imagine manipulating the misfortunes of those affected by the industry. The facts are it's a game of survival for a lot of these agencies. They see big opportunities here now until the 15th of May, hoping that these programs will be extended as far as employment contracts and consulting opportunities like they have never seen around here in the last seven or eight years. There's a lot of it going on. And a lot of people are not benefiting in the industry. I can understand why some people have a bad attitude on account of the way things have been pushed on them.

$750 to $800 million was allocated under this particular program, but $300 to $400 million of that is in the form of direct compensation payments to the fisher people affected. The other opportunity provided here, which is kind of a forced opportunity, is that in addition to receiving a certain level of benefits, you are entitled to options in the area of training and education either inside the fishery or outside the fishery. If people refuse to participate, then benefits could be cut back to the minimum direct compensation which is $225 per week. There are no real incentives to participate in training and education. Had they done it right in the beginning with incentives and a requirement on everyone, the program would have been far more successful. The facts are that about 70 per cent of the people in the industry are only receiving $225 per week because their communities had just about died prior to the imposing of the cod moratorium and their fishing was very limited. This community is one of the exceptions, along with a few others on the Cape shore. We're close to St. John's and we have specific ideas about fishing. Most fishermen in this community are at the top level of compensation, $406 per week. Most of them are experience exempt. And if you are experience exempt, you don't have to opt for any of the training opportunities. If you're at $225.00, the lower end, you don't have to opt. If you're at the top level or in the middle, and you are not experience exempt, then you have to opt for some form of training or you're cut back to

$225. We only have about twenty per cent of all individuals in the province required to do training, and there are some 25,000 that have been identified as eligible both legitimately and through all the cracks in the system. Some of them are not legitimate but they're there anyway. There are 25,000 or more people who are receiving benefits under the program. There's only about twenty per cent of them that are really required to participate in any training and education. That's a gross error as far as I'm concerned. If they were sincere about training and education, there would have been a requirement for everyone, but a requirement that was designed with participation incentives.

Endnote

[1] The Northern Cod Adjustment and Recovery Program (NCARP) ran out on May 15, 1994. It was replaced by the Atlantic Groundfish Strategy—the aid program known as TAGS. Originally envisaged as a five year program, it has been cut back to four with a closing date set for May 15, 1998. Convinced that the stocks will take longer to come back, some fishers have submitted a proposal calling for a ten year revitalization program for the survival of the coastal fisheries.

CHAPTER
3
LOUISBOURG, NOVA SCOTIA

CHAPTER 3
LOUISBOURG, NOVA SCOTIA

ART MacDonald

I worked for ten years at National Sea in maintenance. No pension for me. I don't want to move out of Nova Scotia, so I'll have to do what I can. I'm not too happy about this ten week course at the trade school. But we have to say that we want to take this upgrading anyway if the government is going to top off our unemployment. After 39 weeks of pogey, that's it.

ED WHITE

I joined National Sea at sixteen years of age and I worked in fishmeal for sixteen years. The foreign draggers ruined the fishing. Too many fishermen have been at it. The 200 mile limit should be extended to 350-400 miles. Dragging the ocean floor is like clear cutting a forest. It's the same thing.

41

You goes into a forest with all the mature trees and undergrowth and cuts all the trees down. How many years does it take for them trees to grow back so you can cut them again? I'm not going to see it - and it's the same with the fishery. Twenty years would take it all, if it does come back. The stuff that the fish eat grows on the bottom. When you scrape the bottom, that stuff can't grow back no more. It's uprooted. The draggers work like graders. My grandfather used to fish here years ago, and then the bottom was all hills and hollows. Now you can go out there twenty, thirty, forty, fifty miles, and it's flat as pavement. They go over it so much, it's all leveled. They drag along those big, flat steel doors on the bottom. They go along and they come to a bank and the dragger is going wide open, and it's busted right to hell. And when they drag, they don't go in straight lines. They go back and forth and all over the place.

Another big thing is bycatches. A dragger goes to a certain zone to fish, and they might be allotted 80 to 90,000 pounds of haddock. They are only allowed a certain percentage of bycatches. Suppose they are allowed a 50,000 hake bycatch. If they overcatch that amount and take in 60,000 pounds, they have to dump 10,000 hake over the side. They are all dead because the weight of the nets crushes the fish. That overcatch might be from just one tow. I've seen them bringing in stuff too small to cut. Since 1985 the fish have been getting smaller.

Tracy MacDonald

I'm an electrician now, but I'm affected by the plant closing. When the fish plant isn't working, nobody got no money, the workers don't fix their houses and there's no work for me.

ADRIAN DAIROU

If I take a ten week training course in computer, let's say, how can I get three years education in ten weeks? Who is going to hire me at 50?

This make-work project we are on now is for ten weeks. We get the ten weeks, and then they got to make up their mind whether we qualify. Then after we qualify, we fill out our cards and send them back. In about two weeks, the cards come back. Then we fill them out and we send them back, and they give you one week. And then you start two weeks after that. After thirty-nine weeks, we're on our own.

KEVIN FORD

I worked at National Sea for three years. It is shut down now and it is going to stay that way until the turn of the century. The ten week retraining course is a joke. They're giving you an education so you can spell unemployment.

GERALDINE MACKEIGAN

I worked at National Sea for thirteen to fourteen years. I don't mind working, but I'm not cut out for pick and shovel. My bones ache and my muscles are sore. I could hardly lift a cup of tea last night. At the rate we're working on this project, we might not live to see the fisheries reopen.

EILEEN SKINNER

I worked at National Sea for fifteen years, but I don't qualify for a pension. If you are over 50 years old, they give you seventy per cent of your unemployment over a three years period. I won't be 50 until March. So I'm out in the cold apparently. You had to be 50 when the plant closed. I never worked at pick and shovel in my life. To have to do it at 50 is tough.

ANGUS CAMPBELL

I've been a fisherman for 25 years. Last fall I sold my boat. Overfishing caused the stocks to decline. The big draggers, both Canadian and foreign, caused most of the trouble. About ten to 20 years ago, the trawlers used midwater trawl nets that were miles long. The fish never had a chance. If the fish were not on the bottom, or wherever they were, they could raise the nets. Now they drag the nets on the bottom.

JO ANNE LAHEY

I'm forty one. If they can retrain us for something . . . but what are they going to train us for? I got three kids in school, two at Sydney Academy and one at business college. We've got three travelling. And my husband is trying to get in for CNA, so I have him going to school too. He's trying to get into something that he can have work with. But I don't know. Honest to God, it's really scary. It's a lot of stress. What are we going to do? I don't want to go on welfare. You got your pride, you know. I want to work. We got no pension plan. It's all gone - medical, everything. He's down there twenty one years and nothing now. They are not permanently shut down, so we are not entitled to nothing. They will not give you a permanent layoff because they'll have to pay you a superannuation, and they don't want to pay it.

They're not saying that they are going to reopen, and they are not telling us that they are not going to reopen. Five in management got superannuation. They got a year's severance and that's what we were supposed to get as long as National was not closed down for two years. After two years, they don't have to pay it. Then they're going to release us all, and we're going to get nothing out of it. You get screwed by National first and then by the government. The government is talking about putting millions of dollars into retraining. That's mostly institutions that got that. That's not people going. Once you take the training course, the company and the fisheries people are finished with you.

You imagine my husband went from $400 a week to $200 a week. And myself going to $100 where I was making $250 to $300. We're lucky if we are going to bring in a $1,000 for a month. At least you know you're going to get a bit to eat. Us, we don't know.

CHAPTER
4
NORTH SHORE, NEW BRUNSWICK

CHAPTER 4
NORTH SHORE, NEW BRUNSWICK

LEIGH JAGOE

If they killed some of them seals down there, there'd be more fish. Seals are the ruination of everything. The only mammal that can catch a salmon is a seal. Salmon is their main feed. There's too many seals. They cleaned the salmon nets right out. You can't kill them. It's just like if you shot a moose right now.

Greenpeace stopped it. They used to paint the seals and everything. That's what they done. These people . . . I call them ignorant to a certain extent. All they ever seen was something in a store. If they got out and saw something being killed, they couldn't take it. They weren't brought up to that. They eat bacon every morning all right, but they don't know where it comes from.

I had gillnets one time. I was one of the first ones here setting cod traps. I fished them for two years and I wasted more fish than I sold. And that's the honest truth. I came in and I put them away, and I never used them after. If you got wind here and you got out after two or three days, your nets were full. They were no good for nothing. You'd just let them go and let them drift down the bay. I seen me fire away thousands of pounds. They weren't fit to eat. They were all rotten. One day is all they can stand. And they're still not fit to eat. You catch a cod in a net, and bring it in and put him in the pot, and there's that much brown and brew on top of the water. It's just the same thing as you go and run down a moose or a deer. You overheat him and stop him running. The meat is not good for nothing. It's just like the fish. It struggles for its life. And what's the good. They spoil so quick there. If they were caught right, then you got a fish. If you're using gillnets, you should catch them every day and sometimes twice a day. You can't get there quick enough. And then you have a second class fish. Well, I wouldn't eat them anyway because they are not fit to eat. Any person who is used to fresh fish and then goes to something like that, it just about turns your stomach.

The gillnets that I had were nylon. About 1950, 51 and 52, there was lots of cod. The government allowed these nets to go on the market. We had to buy them. Some of the men got 150 of the monofilament nets. I never had any love for cod-fish. There was an awful waste of fish with the gillnets. I was in lobster and mack-erel all along. I sold my gear this spring.

The draggers are another killer. I was never on one but I seen them here behind the wharf. They were in here catching lobsters. They catch them early in the morning. They were supposed to come in only in separate waters. They had to stay so far off. When dark come, they'd come in and drag for lobsters. That's what they done. The wardens used to go out, but it never made any difference. When dark come, they'd be back in again.

Last spring I was down in Barrachois. We were getting some fishing rings. I was talking to this fellow there and I said, "How about the lobsters?" "A poor year down here, and last year there was nothing." I said, "What's wrong?" He said, "The draggers got them all cleaned out." That's just the words he told me. They just come out and drop the drag, and not only that; it's what they broke and killed you know. Anywhere there's a good bottom, they just drop her and go. They don't do nothing to the big fellow. The bigger he is, the better.

The cod all got caught before they came in. That's the honest truth. The draggers used to line the inside of their nets when they'd go to drag with a smaller mesh. That's what they used to do. They'd kill all the little ones. Once the fish got up on the deck, they were all drowned. They would just throw them over, little cod about that long. I seen that myself. That was on television. If the cod ever come back, I would fish them differently. I wouldn't leave a cod net out or a dragger. It should be all trawls. Trawls is the best. Trawls and handlines. You could make a living on that. Years ago, I seen us go out clearing 1,000 to 1,500 weight of cod. That was in '47 and '48, off Salmon Beach. That was the time we had the salmon nets out. It was in the fall of the year. I seen us come to tide with 2,000 pound of cod there and laying on the handline. I seen Dutch Smith up there one day, and I thought he had 6,000 pound of codfish.

You're wondering why they couldn't keep the draggers out of the bay? They tried it. The government owns the draggers.

WALTER COOMBS

I've seen the cod so plentiful, they would be biting on your hand line and, when they seen you coming, they'd stop. It's true. Before the draggers came, there was all kinds of cod. They tried to keep the draggers out of the bay. They tried. But the government owns the draggers. They tried to put limits on them in Grand Anse, and they couldn't go up any further. But they'd go up this way. There was nothing ever enforced about it, that's for sure.

I think it was last October when the draggers came in at night in the fog to drag for lobsters. Draggers have no license to fish lobsters. Lobsters can only be fished with traps. They were that close off our home there you could tell the colour of the boat. The minute they seen my lights, off. They took off. He was sitting in four or five fathoms. Frank Pettigrew and them went out last fall and they got them with the lobsters right aboard the boat. They said, "We're going to report you." He dumped the lobsters and said, "Report away." They claim that there's nothing they can do if the lobsters are on the deck. If they put them down in the hole, that's different. If they're on the deck, it means they just hauled their drag and they caught them. They could say they are just going to throw them over. They could even put them in boxes and throw them over as they got full. I'm not just talking about the odd lobster. See those green plastic boxes in the back of the truck. They hold 100 pounds. If you got a couple of them full, that's not the odd lobster. Not only that, they're dragging in close and whatever lobsters is there, they're ruining them. That's what they come for, not codfish. You don't catch codfish in three or four fathoms of water. I'm talking about cod draggers, not scallop draggers. The scallop draggers down here don't come too close. They really disturb the bottom though. No doubt they do lots of harm.

HARPER SMITH

Cod fishing in this area is pretty small compared to other places. There was a time when the cod were salted and shipped out, but now 90% of them are all sold locally. As far as I'm concerned, the cod had a major effect on the lobsters. From 1940 to 1960, there were stands of salmon nets off here. Some were 500 fathoms from shore. Well starting about the 20th of May, the codfish would come in and fill those traps. They were chasing the herring and the capelin. They were set for salmon but there was an abundance of cod, and they'd fill their boats. They would be processed in Caraquet.

What I'm coming to is this. In those years we had no lobster. I fished those years with a fisherman, and we fished about 400 traps. Day after day after day you could put your catch under your arm and bring it to shore. In August we would go out handlining. You could catch codfish like that for 7 or 8 months. Well those codfish were eating up the lobsters. Between the escape mechanism which allows small lobster to escape from the traps and the lack of codfish, this is what brought our lobster fishery up in the last ten years. In those years I worked for a fisherman for $1.25 a day, and at the end of the season he pulled in the 400 traps and told me that I had more money than he did. He was fishing for a company. In those times, everything was companies.

I was reading in *Pioneer Settlers of the Bay Chaleur* that, in the 1860's, lobsters were so plentiful that, after a northwest wind, the lobsters would be piled high on the beach. The people would take them by the cart load and use them for fertilizer. Although they were abundant, the demand was limited and the prices were so low that lobsters had no commercial value.

My father was born in 1891 and he told me that there was a factory out here off the banks. And they had rowboats. You know they fished 75 or 100 traps. The little boats fished until the big boat was practically full of lobster right up to the seams. They were three cents a pound. Companies owned everything in those days. There wasn't any such thing as a private license. The company and the fisheries officer got all the licenses. They'd give you a license. It cost $5.70.

In 1933 Frank Hornibrook and Jim Daley were the first two that broke loose and built their own gear. That was 1933 and the company paid them three cents a pound.

In the '30s, the lobster was poor, but there were lots of codfish. The nets were filled with cod. The big 100 foot draggers fished in the gulf twelve months of the year. They were mostly Canadian. That's what stopped the cod from coming.

Salmon were shipped and crated to Boston for eight cents a pound. Moss kept the ice insulated. Everything was shipped by train. It was good for 48 hours. Salmon prices kept going up. Salmon dwindled because of high sea fisheries gillnets. Salmon come home to the same river to spawn. They say that pollution and dams affected the salmon. In my opinion, seals aren't as bad as they say they are.

The 65 foot draggers that are after flounder won't hurt lobsters if they stay where they belong.

As for the future, gillnets and draggers should go. You could make a decent living with longliners and hook and line. Fish farming will not work with certain species. It only takes 18 months for a salmon to reach ten pounds, but a cod takes seven years.

ERNIE SMITH

In this area, groundfish is affecting people, but it plays a minor role. Most of the fishermen in New Brunswick and Prince Edward Island are multi-species license holders. So they can go after herring, mackerel or whatever.

As far as cod goes, it was the mobile technology of the last 50 years that destroyed it. If they come back, we should stick to regulated hook size and stationary gillnets (fixed gear). These are groundfish nets that are anchored at each end. If you go into the hen house and step on all the chickens, you are not going to have a viable operation. We need to return to a passive style of fishing. Longliners can do it effectively with their automatic baiting. A 100 foot boat can fish long line style.

I was vice president of the Eastern Fishermen's Federation (EFF) for three years and this is my second year as president. I know Tom Best. He's a very nice fellow and very active in his association.

I was down in Portland, Maine last year for the World Lobster Conference, and I ran into some guy from Ireland. They were there from Ireland, England, New Zealand, Australia, India — the various countries. They asked me to do a brief on what style of fishing we do in eastern New Brunswick. I just happened to comment on the statistics for our landings in Miller Brook and Stonehaven. It was almost equivalent to the Irish coast. The name of the area is called New Bandon, called after Bandon, Ireland. So some of these fellows came over and they wanted me to go over to Ireland. I said, "Send me the ticket. I'll go." This fellow was in a zoology department. He was from Galway, Ireland.

Groundfish is playing a minor role, dollar and cents wise. In the last eight years, groundfish is probably only playing maybe a ten per cent role financially in our fishing community. It played a big role up until ten years ago when a lot of us quit. It just wasn't feasible anymore. About ten years ago, I changed from catching groundfish to catching mackerel for bait for the groundfish fishermen who were still on the go. A fellow could do just as good at it.

The Acadian Peninsula is the coastline around northeast New Brunswick, Caraquet and Shippagan. Groundfish fishing in the Acadian Peninsula is largely mobile. It's all mid-shore. That's the only part of New Brunswick that's hurting in the groundfish fishery. The groundfish in New Brunswick have been overcaught with present methods of fishing to the point where you're hauling up three tons and keeping one ton and throwing two over the side to die.

We know the migration of our cod. It's all the same cod that comes up here. Back in the '60s, the codfish was starting to disappear a bit, and Romeo LeBlanc at the time stopped the winter fishery in 4VN. That area takes in all the codfish from the Gulf of St. Lawrence, Bay Chaleur and everything like that. Those fish will migrate out to the Cabot Strait in the wintertime, and they concentrate there. So by allowing the winter fishery, with very little effort, they could ruin our fishery. And they were doing that in the '60s, and Romeo LeBlanc put a stop to it. I think it was Ivan Daley who told me that, two or three years after LeBlanc stopped it, that the codfish started coming back again. About ten years ago, they started it again in great style. So the last couple of years, they're cutting out this 4VN fishery again, which is the area between Port aux Basques and Sydney. The winter fishery harmed our fishery a lot here because it's the Gulf of St. Lawrence codfish that migrates out that way for the winter season. And they are in concentrations at that time and can easily be scooped up.

We could go back to a type of fishery which is labour intensive. We're looking for ways to retrain people how to do things. We have tens of thousands of fishermen in the Atlantic region that knows how to catch codfish in the passive longline style. Now let's say they are already trained. You don't have to spend millions of dollars to train them. What we're saying at EFF is that if there is 5,000 tons or less biomass in the Gulf of St. Lawrence, that it be closed to everyone. If it's open from 5,000 to 15,000 biomass, that it only be open to the little guy. In other words, that most people get back into it first. They are paying money to these people not to fish. Get that off of the government costs, and get these people back out on the water. With the under 45' style fishing, the smaller boats that are under 45 feet in length, the conventional style, the passive style fishing - allow them back in. If the biomass, with them fishing, continues to increase to another level, then you allow a more modern, sophisticated style. In other words, the big factory freezer trawlers would only be out there if the biomass allowed it.

I met with the House of Commons Standing Committee on Fisheries in April. I had to go to Moncton for televideo for that. Ron MacDonald is the chairman of that. He's the member of Parliament for Dartmouth. Ron was very excited on that proposal. In fact, when we finished with our brief, Ron said, "I don't want to be prejudiced, but when our report comes out, a lot of your report will be in it." This is an all-parties committee. What they're going to come out with, I don't know. Keep in mind that the tonnages we threw in were as an example. The scientists can determine these tonnages. It's the method we set forth that we'd like to see adopted.

I'm a multi-species license holder, and I fish lobsters May and June. If the fish come back here, I would love to go out there in August, September, October and November and fish codfish. They can keep my unemployment. I'm not interested in it. I would sooner work for it. I would sooner have the codfish that would allow me to do it. I think it can happen.

I've been on draggers at sea and when I've seen what's going on, it would make you cry. Here is what one was caught doing the other year. I've got no proof of this now. It's just hearsay. They had a grinder on board and they took anything that was under a certain size. It went down the shoot and was ground up. That was spit out behind the boat as a pogey. They'd come around the next trip after an hour or two and they'd take the trail where they left the pogey and everything feeding on it, and they'd swoop everything up again.

The inshore is also the vast majority of people. Eighty or ninety per cent of the people are catching ten per cent of the fish and making a living off it. Let them back in first. You're only using ten per cent of the biomass. Where government has failed us is in giving these boats hundreds of thousands of dollars to go to it. They screwed it up. The federal and provincial governments have a conflict of interest in the decision because they've got hundreds of thousands of dollars invested in these boats and, if they go toes up, they're in trouble.

The plants are provincial. Hatfield made the big boo boo on that in New Brunswick here. Gloucester County is the county in the Maritimes with the most fish plants, and it had some 5,000 plus working in fish plants. We just haven't got the fish for it. Anybody who wanted to spend $100,000 of their money, Hatfield gave them a half a million, and away they went. And then the old game was to go bankrupt and buy it back for a firesale price under another name. It was just a racket. They put too many plants in New Brunswick. Lunenburg County is the next place to here for fish plants.

Salmon is still here. Bureaucracy rules. The fellow with the bucks turns around and he goes to the bureaucrats, and the best way to get the commercial out is to say there's none. Many rivers have salmon that are plentiful.

The statistics of what they've taken from the river are suspect. I know that some of those on the rivers told me that they are only allowed to report so much because if they show the river full of salmon, then their enhancement program will get no more money.

I'm clear of the salmon outfit. In the last few years I was getting so frustrated. The last year we fished was the biggest year since they ever fished salmon. The licenses were granted along here by Queen Victoria before there was a Canada, so I don't know what aboriginal rights is. The license was granted along here in the 1840's.

When we were banned from fishing, you couldn't find a political friend. In Ottawa, in Fredericton, I don't care what party, because it was suicide to talk to you. If you're going to support the commercial salmon fishermen, you're going to support the seal hunt. And nobody's doing that. It's not politically correct to support the seal hunt, and it's not politically correct to ask for a commercial salmon fishery. The animal activists are now against boiling lobsters. I was waiting for that for five years. You would find seals in salmon nets. But seals are a lot thicker than they were. The only difference why the seals are in abundance and the codfish is gone is the codfish wasn't cute. If the codfish was as cute as the baby seal, they would have been protected a long time ago. What the loss of the cod has done is to allow the lobsters to triple and quadruple. They're one of the biggest enemies of the lobster. When the lobsters are little like that, the codfish just gulp them. Another reason for the loss of the cod is the government putting millions of dollars into the processing of capelin. That's part of the food chain. I remember the comment Crosbie made. They asked Crosbie one day about the seals. Greenpeace had a full page ad in the Chronicle-Herald. There was some statement about seals not eating cod. They asked Crosbie about his reaction to that. "Well," he said, "I'm darn sure they don't eat Kentucky Fried Chicken."

Frank R. Daley

The draggers caused the decline in the codfish. It's only a small pond of water we got here. This shouldn't even be fished by those people. This is what screwed her up here was the draggers. It's only a small bay eh, and if they soon don't draw the line . . . I realize that they are all government

owned, and the government is behind these things, but if they don't draw the line, we go to bed at night and at six o'clock in the morning they're dragging for lobster around here. They're dragging in here in five, six, seven fathoms of water. They are not even supposed to be in the bay, but they're dragging the bay and they're dragging for lobsters. They're dragging right over our lobster beds. We've been fishing codfish in 18 fathoms and they've been way inside of us. No shots have been fired yet. We like to think we're half civilized around here, but maybe that's what's got to come. I don't think they're allowed in here at all, but they were in last fall. It's hard to control because the government dished out the bucks for these boats and all that. You can't blame the people. If I did not have a job which I did - I worked for 38 years and I'm retired now - I'd probably wouldn't wait to go to the Fisheries Loan Board and maybe get a buck settlement and get into one of these boats. This is what ruined it. This is only a little bay here. You can't stand that. There's nothing being put back in, even with as wonderful a recovery as it has. They just keep sailing along today and, wherever there's a school, they drop their rig and suck them up. Same thing has happened to the salmon. They go to Iceland now. They know where they go in the wintertime. So they're being fished right where they spend the winter. Now there may be the odd one poached, but not very many. One time you could catch them in your nets, but it's only the very odd one now. Now that they know where the salmon hibernate, they go now with these ships and they are being canned and everything right aboard the ship. That's the talk I'm hearing. Something is happening because there's nothing being fished here, and they're declining steadily. Pollution of the rivers doesn't help any either.

I don't know the whole story on cod, but they do have the latest sonar and tracking devices on their boats, and they're caught long before they get to this little bay here.

In my opinion, you'll never see salmon back here again. And how's the cod going to come back? The young guy today who wants to make his livelihood in fishing, I think he's in trouble. And this year the lobsters have declined twenty per cent, and in some cases it's even more than that. The new lobster traps with four doors are catching more. That's something to keep an eye on.

Barry Bezeau

This is my first year fishing and I got 15,000 pounds of lobster this year. The guy I was fishing with we had 18 last year. One boat here last year had 28. The year before, one had 30 and one 29. We're getting $5 a pound for the big ones and $4 for the small. This year we had a good price from the start, but it's not as good as last year. I fish with some four foot traps. They are better, but I would rather fish these. They're not that much better. But some guys fish the square ones. There's an English guy here fishing them, and he's going to take over 20,000 pounds this year. That's all he's got. I had some of those and I don't like them. I find them

awkward. I got 150 small traps from Baie St. Anne, and I think they're better than these. After the lobster season is over, I fish herring. And I got a scallop rig. I don't fish cod. There's none.

We don't know what happened. What's the real thing? I suppose it's the draggers. I don't think it's the seals or the water temperature. The water is always the same here. There was a bit of cod here at one time. When I was young, my father used to jig with just a line, and he used to catch big cod. I remember one time he had one 65 pounds. Usually we used to get some in our traps, but now we don't get anything. We never see them.

A FISHERMAN WHO PREFERRED TO REMAIN ANONYMOUS

What happened to the codfish? At the spring of the year when they are not supposed to fish inside the 25 mile zone, they were at what we call the north west point, and they loaded boat after boat. This was in the spring of the year. And where were the fisheries officers? They were tied up at the wharf at Shippagan.

The draggers filled the shoreline from here to the ice. That's why no fish comes up the bay. That's about it. I'm name shy and camera shy.

Larry Vibert

Well, like myself, I never had that much to do with cod really. It was on a small scale. We did fish them sometime earlier. I never had a longliner. I had a gillnet license, but they took that away from me. They sent me a bill again for it when they said they give

it back to me. I guess the government is low on money now. Last year they sent me back $30 and this year they said if I want my license back to send them the $30. I didn't have to appeal it. I just sent it back because of something the minister said. He decided to give it back because there was too much complaining about it or something. He took it back because I wasn't using it. I wasn't going to bother with mine but, since I had the license, I figured I might as well try to get it back. I don't plan to sell my license because I don't think the cod is going to come back that fast.

I don't think the gillnets affected the cod that much. I think it was the draggers more than anything else - our own boats and the foreign fleets. I never fished on a dragger. I made a trip on one but I was so young I don't remember. They claim that the small fish that came up, by the time they get back in the water, they're all dead. You can't go dragging the fish, and destroy them every time you haul them in, and expect them to keep coming back. There's a limit to everything. Well I think personally, maybe not directly us, but the draggers around here are to blame too.

Maybe something like what happened to the cod could happen to other species. We've seen the lobsters go up last couple of years, but it seems like they peaked last year for here. This year we've started down. It depends on how low it's going to go. We'll only see that in maybe three or four years.

In some ways the big four foot traps are doing more damage than the little ones did. Not in the amount so much of the lobsters they take, but I would say you get your large females into those traps, probably spawners, because I fished them. I got 200 of them. Because you get a bigger lobster in there, we have a bigger entrance ring and well, I'd take lobsters without eggs, but you do get the large females in there sometimes. I talked to another fisherman about this a couple of years ago, and he said, "Well, what's the difference whether we take them or somebody else takes them." But if you're looking to the future, we could be destroying ourselves, our livelihood. I don't know. I got one the other day with just the two claws when he was caught in there, eight pounds and a half in one of them four foot wide traps. The thumb on his pressure claw was sticking straight up. He got it in there but it stopped him from coming back out. He was hooked on the other side.

I think I've reached the limit of all I want of them big traps. The big traps aren't catching more lobsters on a regular basis. If you can find like a spot of lobsters, you'll do well with them. We didn't have no great lobster spots this year. Last year I did, and I found my quota of lobsters were plentiful. And if you set them there, you'll catch more there. This year we never found no big amount of lobsters any-where, and it was just about the same all over, and these traps were just about equal. When you got lobsters to work in and you've got the room for them, then you'll get more. If you only get there every second day, it's better.

I seen one guy the other day. He caught a cod fish. He was alongside me. The codfish was as pitiful as I've seen this year. It was skinny. Usually, if you get a cod here in the spring, and there is capelin around, they had a nice belly. This here was streamlined.

DOUG VIBERT

So you were talking to Larry. Larry's father is a first cousin of mine. Overfishing ruined the cod. The first trip I went aboard of a dragger, I saw right off the bat that that was impossible to last. When you hauled the net back, you saved a thousand pounds, and a couple of thousand pounds went overboard and got lost. That's it. I don't have to tell you anymore. They were all dead, floating on the water. And all the boats doing the same thing.

I don't know what will happen to other species. Lobster is something I can't figure. But they're dropping this year because I know for the last 50 years, the people of Miscou have tried their damnest to destroy the lobsters. They don't know it. They're getting bigger traps. If they would stay home on Sundays, just look at how many lobsters would be left on the ground for another year. But they have to take Sundays. Here there are around 60 boats. If they only get a couple of hundred pounds a day on Sunday, you can figure out the amount. A fellow from Nova Scotia went fishing on Sunday. They took him to court and all this. They were going to take his license, but he got away with it. Now it's just another day. I never did it.

The bigger traps fish better. There's more lobsters goes in them. If we were fishing the same kind of trap that we fished all our lives, then there'd be that many more lobsters in the water. That's the way I see it. The last three years around here have been the best years we've ever known around Miscou for lobster and for price. They're getting 5 and 4.

If I were the minister of fisheries, I would cut out Sunday fishing.

Codfish is not going to hurt Miscou because this wasn't our main industry here. Years ago, people fished a few cod but, for the last 30 years, nobody fished cod on Miscou.

I'm enjoying my retirement. I've been fishing since I was ten years old, so I've done my share. You know what they say about old fishermen. They never die, they just smell that way.

DONAT LACROIX

My father was a fisherman, and my grandfather too. My father didn't want us to fish because he used to fish with my grandfather aboard schooners. It was very hard. So he sent us to college. After we were through college and university, I worked with the Department of Fisheries. I quit all that and I went fishing. I've got two boats.

When I was in college, I performed in theatre in 25 to 30 plays. Today I write my own songs. Most of my songs are about the sea. The dreams of the fishermen, some lament songs, and stuff like that. One LP, three 45's, one CD and one cassette.

When I finished my BA, the old classic course with lots of Latin and Greek, I had to choose a profession. When I worked at the Department of Fisheries, I had jobs in the summertime. I had to go aboard draggers. I was on several draggers. And once I was surprised, aboard a dragger for three days or five days, at the amount of small fish that they were throwing back. You can ask my wife. I said to her a long time ago, "This won't last long. We are going to suffer for that, for the pillage that we're doing. We cleaned out the babies." I talked to a fisherman several years ago. He was from Caraquet, and he was fishing off Cheticamp. In springtime they go there. He said, "I caught 8,900 pounds of cod, and I saved 900 pounds. Believe me what I tell you. And if I had to save those over 18 inches, I would have saved 200 pounds. And there were many boats like mine there."

Now, who's to blame? I have often heard fishermen complaining, but I never heard them complaining to sell. I talked to a fisherman once. I asked him if he used a lining. You know what an otter trawl looks like. There are regulations about the size of the mesh to let the little fish get out. He admitted that he did use a lining and, when I asked him why, he said, "Everybody does it."

With a liner you catch everything. If you get caught with a liner, you should lose your license forever. If the government made a regulation that you don't process under a certain size, say 17 inches, the fishermen would applaud that. It would allow the small fish to grow and the females to reproduce. But there are pressure groups and workers who want work, and all these attack. So you process small ones. There's no reasonable size fixed by law - let's say under 21 inches. That would have helped.

Some lobster fishermen keep females with eggs. You shouldn't. It's very severe, severe, severe. There are about 50 lobster fishermen in Caraquet. The Department of Fisheries tagged 1,600 lobsters last year. And they dropped them in the Caraquet area. They wanted to see how far they travel. The officer was thinking that we might catch around 200-250 of those tagged lobsters. I caught thirty myself. I've got four in my frigidaire there. He's supposed to pick them up. We have to take the number of the tag, the depth of the water, the position, the longitude and latitude, and the date. We'll catch more than half of those tagged lobsters up here. I'm anxious to see.

Now the limit of traps here is 370 traps per boat in Zone 23. We used to fish with wooden traps. This year I experimented with the wire mesh traps. They are four foot traps with a double parlour. I almost doubled my catch.

I'm afraid we went wrong. In the south it started near Buctouche, and the lobster catches there have been going down. For me, the fisheries department could have a closer look to that, and those big four foot traps. With 36 inches, you cannot have two parlours. The big ones are almost like two traps in one. I'm afraid in seven or eight years from now it could go down. The biologists claim that 20 per cent of the lobsters that come escape. I'm afraid with those traps we'll bite on that 20 per cent. In my opinion, we need to take a close look at that before it's too late.

For Caraquet, cod doesn't affect us much because here there is crab, herring, lobster and shrimp. Probably five to ten per cent of the production here in Caraquet is groundfish. When there were more than 40 big seiners here from Nova Scotia fishing herring in the area of Caraquet, they got herring. Now there are only four herring seiners in Caraquet and, for a while, they cut their quota quite low. And licenses that you couldn't buy because you didn't have enough money . . . at a certain time you will cheat because herring went down and inshore fishermen were allowed to fill their boat. I filled my boat with 80 barrels a day, five days a week. Eighty barrels! My boat can hold 200. There's a quota on my license. On my license, you should not take more than so many fathoms of seiner net and 80 barrels a day maximum. I did that until we had no more herring. We can learn a lesson from that.

The crab almost got it too. Now each fisherman has a quota for the season. There are shore inspectors. Let's say I have a quota of 150 tons of crab. When I reached that, that's it. I'll probably have another quota, bigger or smaller, next year. It depends.

The herring quota is so small, it's not worth fishing it at $10 maybe $15 dollars a barrel only for 850 barrels. We got expenses.

We need twelve stamps for UIC. The maximum you get for the lobster season is ten stamps. Some years nine. You need two more. So they have to fish away. Not for the money.

There are a lot of pressures. Union workers and shop workers want jobs. They want work. To work means fish, and the fish is in the sea. Fishermen bring the fish.

The amount of fish the inshore fishermen take of the total catch is very small. And we are the first ones to lose our licenses. I lost mine because I need a receipt to show for the years 1990, 91 and 92. Why? Because I had a longliner trawler and there were no more cod. I didn't destroy the cod. The draggers did. I had no invoice to show. I could not sell. I could not fish. There was none.

CHAPTER
5
SOUTH EASTERN PRINCE EDWARD ISLAND

Chapter 5

South Eastern Prince Edward Island

Danny King

The lobster season is in May and June. After that we go after herring, tuna or scallops. In their latest decision, the government of P.E.I. is limiting tuna licenses to 50 boats. No one wants to lose a license. Everybody is trying to use every license they have. Once the lobster season is over, the only thing left at that time is groundfish, because the scallop season doesn't open until October 1st. We are losing money in trying to make enough money.

Factory freezer trawlers fishing in the Gulf of St. Lawrence finished the cod in this area. Now the big 100 foot draggers are fishing for redfish, and they might often have ten to 20,000 pounds of cod on board. They are allowed a ten per cent bycatch. But individual fishermen are not allowed to handline for cod!

FPI and National Sea each had an excessive number of boats. If a captain found a good spot, he alerted the other ships working for the same company and they cleaned it out. That's not fair competition. They were there for profit. They were not there for long-term livelihood.

JACK KING

The lobster catch has been going down here for the last four or five years. In the '30s, we used fifty trap lines. Now they have dog trawls or bull trawls with four to ten trap lines. They can be moved every day. There's no room left. 300 traps are allowed. Limited entry means that the only way you can get into fishing is to buy an existing license. I know that the lobster fishery is closely watched with limited entry, but the area is too big to manage properly. There are too few officers for too big an area.

Once the lobster season is over, you might go after hake and cod. At the start you might get 600 pounds a day. As soon as the draggers are allowed in to start fishing, the catch for the net fishermen drops to 200 pounds a day or less.

If you are caught with one salmon in your boat, they can seize your boat. I wish there was a booklet for all fishermen outlining all the regulations. If they wanted to make changes, a three ring binder or some such thing could easily be supplemented each year. Regulations help, but mind you there's often a big difference between what the fisheries officer says is an acceptable bycatch and what is actually unloaded.

The big fish companies have people lobbying for them. In the last few years, they've modified the drags for groundfish with rubber discs. We call them rock hoppers. They do a better job of sweeping the bottom even in uneven terrain. It ploughs up the bottom. Even hake is getting scarce.

An individual is allowed 300 bar clams a day. That's too much. People are making money selling them on the side. A five gallon bucket full would be a more reasonable quota.

The bigger traps catch more lobsters, and that could knock it back a bit too. If you get a bigger trap, you can fit 15 or 20 lobsters into it, no problem. If you get a small trap with five or six in it, it looks like it's full. You're catching three times as many in the big traps. One trap is one trap. The other guy got 300 and you've got 300 big ones, and it's like you got 600. There's a limit on the trap size now and they can't go any bigger. But still, they're big enough. I don't think it makes sense for the government to license Clearwater to convert a dragger into a fishing boat to go out after the big twenty pound lobsters. The big breeding stock - that's what keeps us inshore fishermen going. You can fish lobsters farther out. As long as there's rock on the bottom, you'll get them. Especially in the winter time like out in the ocean farther where the water is deeper and there's no ice. The deeper the water, the warmer it is in the winter time. So that's where the lobsters are - in the deep water in the winter.

The government is making up these laws, and they really don't know what they're talking about because they are not out there first hand doing it. They send out a scientist and he catches a few lobsters to see how the stocks are, but you can't go by that. You've got to ask those fellows who are working at it all their lives.

The offshore fishing is a mess. You've got these big factory freezers. They're out there for months loading up. The inshore fishermen are selling to those guys. They are just stocking up solid, even the red fish and every other kind of fish going.

They should have cut back on those big ships 20 years ago. If they had done it, there still would have been plenty of cod for the inshore fishermen right now. For years the cod were caught before they got near the inshore.

It doesn't look good, you know. Probably next year I'll be taking over the lobster fishing. It's not a good looking industry really to look ahead, if it's not going to be looked after.

The different sizes for lobsters is causing problems. There are fellows that buy on the north side, and here some guys are taking in the smaller lobsters than we can take and selling them to the trucks that are from that side because they already had the shorter lobsters on the truck. If they get caught with them in the boat, they're done. But as long as they get them on that truck, because they are a legal size, they got them from the other side of the island. The lobster fishermen are just cutting their own throats, and they're cutting everyone else's throat because they're taking undersized lobsters. It would kill me to take an undersize lobster. There are laws but they might not be good laws. If they are going to make a law, make it for the whole island. It's not that big an area. But I guess over there they don't get many big lobsters. We get a lot more markets and stuff like that, and all they get over there is lobsters that make the gauge more or less, and small canners.

The future is just a gamble. The factory freezer ships take enough in one day to keep you going for a season. With all the latest equipment, the fish don't have a chance. Are they just going for the short time thing? Do they have any idea of what's going to be down the road for them or what? There's nothing there now. What are they going to be doing?

They built three draggers for Usen Fisheries here in the shipyard when I was working there a couple of years ago. I forget what they cost — about $5 million. There's hundreds of people down there that are out of work. The plant is gone. The quota is gone.

Georgetown is a fishing town. If there's no fish here, the fishermen are afraid and the people in town are afraid because there's no work in the factories.

I might have to get into a new trade. The government doesn't want you to fish cod and stuff like that, but if you don't fish for them, they'll take your license. They are contradicting themselves. I don't think they realize what they are saying. They are just saying it to please other people.

Doug Sorrie

I haven't bothered with cod for ten years now. I used to drag, but now I just lobster fish and scallop.

It's hard to believe the amount of seals that is around here now. The guy that's making the most money that I know of is taking people down to see the seals.

The hake is gone from here, wherever they've gone. A lot of boats from Murray Harbour south used to drag and they would come in with boat loads. Now there's nobody dragging because there's just nothing there.

My wife Myrna and I fish together. We get up at 3:30 in the morning and we're finished around dinner time. You generally have your best weather early in the morning. If it starts to blow, it is generally 8 or 9 o'clock, and by then you're pretty well done. That's mainly why people go early. Myrna says it's because they can't wait to see what's in the traps.

The lobster fishing has gone back for a few years. I've been fishing 28 years. Up until ten years ago, 8,000 to 10,000 was a big year. For the last few years, we've been up to 20 to 25,000. A lot of people are down from 10 to 15 this year. I think it goes in cycles. The fellows at East Point had an awful poor year and up at North Lake, only ten miles away, they had a good year.

The 30 crab fishermen in P.E.I. are on a quota of 75,000 pounds per boat per season. At $2.25 a pound, that's not bad for a three week season. The crab fishermen here are fishing lobster at the same time. They'll go out at night and haul up crab traps, and maybe they'll leave their lobster traps for two days. It's changed around some. Some of the guys sold to different guys.

The loss of the cod has a lot to do with the big boats. They've got such good equipment now that they know where they are and what kind. What we would get in a whole summer when we were dragging, one of those boats would get in a few hours. Right in our own 200 mile zone, there's a lot of foreign boats fishing. They have so much quota and they have to allow them to it. Our own Canadian boats can't get out and fish. That situation is not helping matters.

As far as the big traps go, the traps themselves aren't the problem. It's the ring size that matters. Now they range from 4½ to 9 or 10. You're going to catch an awful lot of bigger lobsters with 9 inch rings. I think they should make the ring smaller. That would protect those great big females with millions of eggs on them.

It's aggravating to have three different sizes for lobster. Maybe votes count or political influence or a lot more hollering. There was an awful fight on here last year. The fellows were dragging with the rock hoppers right over the top of the herring spawn. They weren't giving even the herring a chance to spawn. Fisherman's Bank, out here, is one of the places where they fish herring. For the last two years they've been using the rock hoppers and dragging right over one of the places where the herring spawn right on the bottom. Fishermen are greedy too.

When they started out at first, you couldn't make more than twenty five percent of your living other than fishing. That was putting more strain on the fishery. People could go and work, and they would say, "No, if you're going to work, you won't be able to fish." They cut you down to 90 traps, and they had other jobs. We were fishing 300.

We used to have all kinds of haddock up this way. Then they built the Canso Causeway and there was never any haddock after. They used to come in there in the warm water and they won't go up around the north in the cold water. No one predicted that would happen. The haddock stay in the warmer water. They get lots of haddock on the south shore of Nova Scotia, but it ruined it for here.

Edwin McKie

The P.E.I. Fishermen's Association is represented in almost every harbour. These are people who paid their dues. Some years ago, the federal ministry gave out some crab permits to 30 people which the Fishermen's Association lobbied and looked for. But when they got their bag of cookies, they went behind closed doors and formed the Crab Fishermen's Association (CFA).

The government of P.E.I. decided that we needed a better fleet of boats to go after groundfish. They subsidized a group of individuals on P.E.I. that always had lobster licenses and scallop licenses and groundfish licenses. They got their bigger boats and decks and more horsepower and started looking for a bigger share of the pie. Then they went behind closed doors and formed the Groundfish Fishermen's Association (GFA). The only thing about these two associations is all the people that belong to them also fish lobsters, tuna and scallops. They tried to lobby for the best for their guys. The majority of the groundfish fishermen are in the P.E.I. Groundfish Fishermen's Association (PEIGFA). The only ones that aren't are the ones that have middle distance boats. They don't fish any farther offshore, but they fish further from home.

There are six regions that make up the P.E.I. Fishermen's Association. Once a month 3 or 4 representatives from each region meet in Charlottetown. The PEIFA has an office and staff to help things along. It probably won't work in Nova Scotia or New Brunswick because of the distances people would have to travel. An hour and a half is the farthest anyone lives from Charlottetown. The PEIFA has been working now for close to twenty years. The majority of licenses from P.E.I. are multi-species - lobsters, scallops, herring, mackerel and tuna. There are about 860 people on P.E.I. holding groundfish licenses.

There are 1,400 bona fide fishermen on P.E.I. and 1,306 are lobster fishermen. The breakdown on the exact figures is in the Cashin Report.

The average catch of lobsters two years ago on the whole island was 17,000 pounds. This year the average is probably 12,000 pounds. It was as high three or four years ago as it was since lobster fishing started. Some people are concerned about the drop, and maybe they should be, but it's unrealistic to think they would continue as high as they were.

What happened to the cod happened to the hake here six or seven years ago. At this end of the island, the hake was far more important from here to Murray Harbour and all along the Pictou Shore and Antigonish. Hake is frozen and filleted and salted. There's a lot of salt hake shipped to the Caribbean. Hake went from about 19,000 tons to about 3,000 in the southern Gulf. When I started otter trawling, my father and another fellow were gillnetting. There were summers that they hit 100,000 pounds of hake in six or seven weeks of gillnetting. Now there wouldn't be that much landed in the 2,000 put together. No one knows the exact reason for the drop in hake, but we always thought there was too much pressure put on them on their spawning grounds. When the hake came into the southern Gulf to spawn in the last week of June and the first couple of weeks of July, the middle distance boats really fished them pretty hard on the spawning grounds.

I fished a dragger for National Sea. The offshore boats definitely had more to do with the decline of the codfish. There was an application made to close the offshore fishing in 4VN. That's the district of Cape Breton where the codfish winter. Gaston Godin from APPA, a 65 footer fishermen's group in New Brunswick, put that together. He made a real good case for the closure. It was called The Request for the Closure of the Winter Fishery. That was about four years ago. The P.E.I. Department of Fisheries and the Nova Scotia Provincial Department of Fisheries and the offshore lobbied to have it kept open, and they won. That year they went 6,000 tons over their quota. The Department of Fisheries said it's not really a real problem because the inshore fishermen didn't catch theirs. The first thing you know, the Gulf was closed down completely and the quota slashed from 43,000 tons to 15,000 tons to what it was last year.

I don't think we should ever have stopped the small boats with fixed gear and handline. I think they should be out there fishing. The amount they take is minimal, but it's an indication of the state the stock is in. And the handliner or the guy jigging over the side of the boat, if he definitely isn't catching something, he's going to stop. Biologists can make tests and predictions and crunch numbers, but if you have boats scattered out from Cape Breton Island, through here and up the shore of New Brunswick to the Gaspe, getting fish in a reasonable number of 200 to 300 pounds a day, it would give an indication of what state the stock was in, and you would see patterns as to where the fish are and how many there are.

Let's say a fisherman catches 600 pounds a day at 50 cents a pound. It doesn't seem like a lot of money, but if you spread that over the summer, four or five days a week, and your expenses are almost nil! If you take the whole fleet of boats out of North West Harbour and put them out there handlining, it wouldn't cost as much as it would to fire up one National Sea trawler.

When the codfish stocks came up in the '60s and '70s, I think the 1977 allocation for the Gulf here was about 15,000 tons. By 1984, it was up to 60,000 tons. At 60,000 tons of fish, there is a place for bigger boats. The resource is there and we can't have it die of natural causes. The fishery has always been used for socio-economic reasons. It's a good idea. So if we're going to do the best for the most, then we let all the little people in with the small boats, and spread it around through the community. As it gets better, then you can start putting the bigger boats in.

If you don't do something to protect individuals - and we have people on P.E.I. with $500,000 boats - they are out there killing fish to make profits for ALCAN, the aluminum boats or the big fibre-glass boats. When it's spread around, the crunch isn't as bad for us with other fields to fall in.

With the technology we have today, we have 65 foot vessels whose fishermen can land as good a quality fish as a 35 foot boat. The problem we're faced with is as the technology gets better, the bigger 45 foot boats are displacing the 35 foot boats, and the 65 foot boats fishing in the normal area are keeping the fish cleaned up before they get in to the little guy. To do it in a fair manner, the quota should be set up so the inshore gets theirs, and then it's farther out.

One of the other things we're faced with is individual transferrable quotas. Boats get into a situation where vessels are given individual transferrable quotas based on their history - and you've got to wonder how they filled their history up - for their past four or five years' landings. The Department of Fisheries used the weigh slips that these people brought in. But at the same time they instituted a Weigh Master system because they didn't believe their own figures. And the boats have to pay for weigh masters, for people to stand on the wharf and watch every fish coming out of the boat. So to get individual transferrable quotas, people will buy and lease somebody else's fish, and we have general figures now where, should the fish be back in the fisheries, we'll have half a dozen people in this community standing on the wharf trying to sell their lease or quota to one of the other people, if they want to make a living. It cuts down on the number of people in the fishery, but does somebody need this fish to have a viable operation.

All the blame doesn't go on the offshore. Every guy kills fish. And the government has a real conflict of interest. They finance the boats, they insure the boats, they allocate the fish to the boats and they subsidize the fish plant.

There are things like red fish and silver hake. You need bigger boats to catch them.

After the Kirby report, things got into full swing. In 55, when National Sea was restructured, they were having a whale of a time spending thirty five cent dollars. Expansion was the buzz word. Bigger boats and bigger fish plants.

The stocks could come back reasonably fast with the right steps. You can't be going out with 4¼ inch mesh and expect to have a fishery. If the ground fish had been managed with the same respect as the lobster fishery, this never would have happened. There were times here at Souris and Cheticamp where there would be 600 to 800 codfish in a thousand pound tow. A trout fisherman would throw it back. The fish would have to be a pound and a little bit apiece. If the fish were five pounds apiece, you'd get 200 fish in a tow. So that was part of the problem.

I start off scalloping in the spring, then the lobster fish, then the gillnets in for a short time and, if there was some mackerel around, I'd be mackerel fishing, purse seining that is, but there's not many mackerel around. I used to gillnet for herring when the price was decent. I think if we had been taking better care of the herring earlier, we could have developed something earlier. When the fishery for herring roe started here, everyone just shook their nets into the boat, took them in and pumped the fish out. The fish weren't iced or anything like that. Gordon Cummings, when he worked at National Sea Products, said you can't send fish up a pipe at one hundred miles an hour and expect to do great things with it. That's when we really should have been working at NRC to be doing something with the carcass. We would still have a herring fishery today. If the price of roe is down but you're taking another nickel out of the carcass, you'd be all right.

There's a lot of bones in herring, but the vinegar solution dissolves them. There's a fish plant in northern New Brunswick where they fillet the herring, salt it and barrel it and ship it to the States. It's made into Solomon Gundi. There's a little bit at Mercy Seafoods down in Liverpool. The biggest thing is to try and get a share of the market. Getting the products on the store shelf is probably as big a job as catching the fish.

BERT BOERTIEN

Those big boats killed off the cod. They caught 50 to 60,000 pounds at a time. Now they are going after the redfish with small mesh gear. But this area has always had plenty of cod and, if the moratorium lasts a few years, I think they'll come back strong. Just lately the DFO ran a test and over 60% were cod. There are also reports of spotting huge schools of cod.

Some of the scientists were saying that the cod are so scarce that they come close to shore looking for food and go right into lobster traps looking for dead bait. That doesn't make much sense when there are all kinds of herring for the cod to eat.

Ten thousand tons of cod were caught in the recreation fishery. Brian Tobin (Federal Fisheries Minister) had to shut her down. The fishermen were peddling them all over the place.

In the early '70s, you couldn't catch a codfish here. There was a two cent subsidy to catch them, and the plant got a six cent subsidy to process them. You couldn't get a thousand pounds for a whole day's fishing. I've always dragged for fish.

Before the 200 mile limit came in, Danish boats fished alongside us with small meshes. That helped kill off the stocks. Most of the inshore fishermen from the Atlantic provinces all have the same idea - kick all them big boats out. Romeo LeBlanc tried to do something, but he was put out in a change of politics. I was president of the Eastern Kings Fishermen's Association for years.

The 45 foot boats create more jobs and they are just as effective as the big trawlers. They are a lot more productive for the country. I'd like to see the big draggers go. We would be all right with longline and gillnets, and the small boats could easily convert from lobster to scallop.

I hear that they've had a better year than they've had in a long time for lobsters in Newfoundland. That would be more on the west coast.

Around here the lobsters went down the last three years to about 15 or 20. About three years ago, they were getting 30 to 35 around Souris and Murray River. I don't think the big traps should be allowed. They've gone from three foot to four foot with double ends. The limit is 300 traps. By adding the extra end on the trap, you add one half to your total catch.

The fishermen here asked the DFO for a regulation on the ring size. There is an unwritten agreement here that we limit the ring size to five inches. Around Pictou and River John, the fishermen have been having a terribly slack season. They have been complaining about the ring size. Fishermen can be greedy.

Three or four years ago they introduced the big traps. Since then the catch is going down.

Hake was numerous. A few years ago we caught seven million pounds of hake in this harbour. The 45 footers cleaned up the main spawning ground between back of Pictou Island and Lismore.

Baie Verte was another spawning ground for hake. They used to make their living on hake. Now they are all gone. All because the DFO wouldn't listen. We asked them to close the areas up during the spawning period from July 15 to August 15.

The MacDonald Commission proposed to get rid of the inshore fishermen. All in the name of efficiency. Who would profit from that?

The fish plant in Souris burned a year ago. All the fishermen, all the plant workers and those in the area not directly involved in fishing wanted the government to keep the quota in Souris. The DFO is still dragging their feet on that one, and the plant is not rebuilt yet.

The Newfoundlanders screamed that the big boats were gobbling up the fish. Scientists said that the codfish were moving offshore because the water is getting colder. Bullshit. Those scientists weren't fired. They are still on the job.

The capelin were overfished by foreign and domestic fleets. They were sent to Japan. They bought the whole fish for the roe. They invented a machine for dividing up the fish. I saw it on the boats. The males would fall through the slots on a moving belt and the females would stay up. A protrusion on their sides kept them from falling through. It didn't take long to decimate the capelin. And we're still sending seiners after a few capelin, trying to catch the last one.

I've been at the wharf when a small boat came in with about one hundred fish, and the DFO officer would count every one. Then Usen Fisheries would come in and they wouldn't even count them.

The fishermen in Gaspe were complaining that the big draggers going after redfish were tossing over thousands of redfish because they were too small. They were all dead of course. Clark, from Newfoundland, who was head of the Conservation Council, replied that perhaps there should be more observers on board. Now I know there are some observers on foreign boats, but there are none on domestic boats yet. They all talk about observers, but they don't do anything about it because that would mean stepping on the toes of the big fellows.

EBER WILLIAMS

Back in 1976, I got the idea of doing what the Bluenose done. My boat, the Cape Colleen, was 46 feet long. I put the salting aboard — salt banking they called it. I left home Monday morning. There was lots of fish around then. I had my long lines aboard, trawls, you know. I had 1,000 hooks or so and I put out six mackerel nets. I was all alone and I put a nice boat mooring out, half inch rope and a big piece of chain and a big anchor. Twelve miles off Arisaig Chapel, and eight miles east of Pictou Island. So I stayed there seven weeks. I'd get up in the morning and I'd haul my mackerel nets, and I'd get about 75 pounds of mackerel and cut them up and bait up my long line, my trawls, and I'd set it. At night, at 9 o'clock just before dark, I'd go back to my boat and set my trawl from where I was anchored. Then I'd go back to my boat mooring and I'd shave and clean up for the night, listen to the radio for a while and go to bed. I'd get up in the morning and go and haul my trawl. I'd dress my fish, split my fish, salt my fish. I was carrying 30 or 40 bags of salt all the time. I was salting the cod and the hake. Every day I would follow the same procedure. I'd come in every Saturday night. When you left Monday morning to go, you couldn't unload the catch for the first week. They had to be pickled for fourteen days. When you came back in the next Saturday night, the tanks were ready to sell. So you cleaned them out Monday morning, sold them to the fish plant and you put aboard your 30 or 40 bags of salt and you took off again. You were unloading fish every Monday morning after the first Monday.

We were baiting trawls and going out, and we were catching fish and coming in and salting them in our fish houses. But there was a lot of running back and forth for fuel, and you wouldn't have your own mackerel bait because you'd have to buy it from the factory and because you wouldn't be out there to haul your mackerel net.

So I stayed right on the ground like the old salt bankers. The fish were . . . . 55 cents a pound and I was running 200 pounds out of pickle a day. After I was there the first week, I was up to 400. I was making $200 a day and no expenses. And I was getting six to eight weeks at $1,200 a week. Salt was cheap then, at 90 cents a bag, and the government gave you half of it back. They subsidized you for your salt. I was carrying 30 bags. A pretty small bill for salt. It worked out real good. If I was a young man, I'd love to do it again. But there's no fish now. You wouldn't make $100 in a month.

I pair seined. Two boats towing a Scottish seine. You shoot away your rope, then you turn on your cable, and you come down, and the other boat shoots away and goes that way and he comes down, and you tow these big ropes and cables. Then you come together and tie up, and then you put your winching gear on and you haul her back. We got some awful loads of fish back in '70, '71 and '72 and '73. In '74, '75 and '76, I went to the trawls because we cleaned her dry.

The Danish seiners have the other boat riding on the back of the main boat. It draws the fish to the net instead of taking the net to the fish. Maybe 8 to 10,000 pounds in a haul.

I know a Captain George Thompson from Glossymouth on the north coast of Scotland. He was about seventy years old then. In 1973 he said, "I'm going to tell you something. If you fellows keep towing these nets around and dragging all this rope over the bottom, in twenty years time the government will have to close her down." And last year in '93, Crosbie says, "We've got to close her." We could load a boat with one tow. We could take our two boats, 46½ foot boats. We could run that net out and tow up here and come together and haul them in. You would have 15,000 in one tow. That would load a boat. We'd tow about three quarters of an hour and load her and, the next try, we'd make another three quarters of an hour tow and load the other boat. We'd be in in the morning perhaps at ten o'clock. Now they're out there dragging the net around all day and you might get a bucketful.

When we were pair seining, we'd go after whatever we could get - cod, hake, flounder. The big factory freezers damaged the northern cod stocks, but we didn't help it either.

The scallops in the strait are scarce now. In the fall we've got 200 boats out there tearing around. At $9 a pound, you might catch a hundred pounds some days if you're lucky.

Back in the '60s when I was at it, you would have 200 pound of scallops aboard by dinner time. They were plentiful. Price will make up for it.

One thing that will look after itself is the lobsters. You're only allowed 300 traps, and there's no new licenses. If there are six boats fishing here this year on the grounds, there'll be six next year.

The big thing around here lately is the tuna fishery. Down home last year, we had fellows getting 8 or 9 tuna at 800 pounds each and getting as high as $42.50 a pound. One guy got two a day for three days in a row. Six fish and they averaged over $20,000. He made $120,000 in one week. Of course, there's another 150 boats that never saw a fish.

There's a fellow down in Canso who began fishing tuna in 1971. He fished every summer and every fall, and he caught his first tuna in '89, 18 years later. Not much money in that.

I hope the big ships stay right where they are. I hope the big ships clean up the cod. I hope the cod never get back, myself. My dad told me in 58, "You'll never have any satisfaction with your lobster industry until the draggers clean them out of the Northumberland Strait. As soon as the codfish are cleared up, you'll get more lobsters." He had sliced open a big cod years ago to see what they were feeding on, and twelve or fifteen lobsters ran out on the drag. My father was fishing with me at the time.

Four or five years ago there wasn't a codfish here in the strait. 30 or 40,000 pounds of lobster to every boat. When we were fishing them back in 58, we got 4 or 5,000.

When the lobsters shed their shell, the cod gulps them down. Up around the north side of Newfoundland, where there's no lobster fishing, it's all northern cod. The Labrador coast and way up farther, I'd like for the cod to come back for these guys, but I'd never want to see the cod come out here. You can't have both. We used to make $10,000 on the lobster and $30,000 on the cod.

There's no doubt that the big factory freezers gobbled up the cod. But there's another way of looking at it. How many of these big boats are out on the Grand Banks tonight? There would be a couple of hundred - but no Canadian ones. Are we crazy? Why let the Norwegians and boats from Panama and the Spanish trawlers scrape her dry? Why didn't we hang in there and scrape her dry with them? If ten fellows are dragging, and one's not, is the one fellow right? It's no good for us to be in while they're still at it. Why don't the other fellows come along with us? I'd love to see them all tied up. But I hate the way we're doing it. All our draggers tied up and Ottawa feeding our fishermen with this here little welfare scheme every week. And all the Norwegians and the Portuguese and them fellows out there with their factory ships putting up the loveliest fish. They're going to be on the Nose and Tail and everywhere else at night when there's no cutters around. What's on the Nose and Tail in August, in November is on the Norwegian ground. They swim you know. They don't stay in place. There's no good protecting the Nose and Tail of the Banks when something is going to be swimming 100 miles every month. They go to the Gulf Stream in the winter. Fish are on the Nose and Tail for two months in the fall. The other ten months, they're somewhere else. Everybody else is dragging but us. I can stay home from lobster fishing every day all season. It wouldn't help the lobsters any if my neighbours are going to haul their traps. They're making a fool of us, man. Ottawa is paying these fishermen so much a week to stay home so Portugal will have good catches. I don't know what Crosbie was thinking about. Pull our boats to shore and start sending out warships to fire shells across the bows of the fellows that's out there. That would do a hell of a lot of good to your fishery. Good conservation in that, shooting at boats that are dragging. Then he said his army wasn't big enough to send any warships out. "We better forget that part," he said.

If the fish do build up, they are not going to stay within the 200 mile limit. They'll soon swim across the line where the other guys are.

Bruce MacNeil

We really need a cod fishery in Atlantic Canada. It's important to our economy to have a balanced fishery. Our lobsters dropped down thirty per cent this year. You can't blame that on the cod. But definitely, if you have more large cod around, they're going to eat more lobster. Cod eat everything. They eat a lot wider range of stuff than a mackerel would. Mackerel is not a bottom fish.

There's probably as many cod around here as there ever was. That's not the main school of cod that's in the Gulf, or northern cod. The northern cod, in fact, is a different stock. Generally, we only get just a scattering of cod up to the strait here. We don't get big migrations of cod. The big migration of cod passes north of P.E.I., down towards Cape Breton and down towards Cape North. It seems like when the cod get up in the Gulf so far, they scatter all over the areas, and we get a scattering up to the strait here.

If we were dragging ten years ago and had 4,000 pounds, you'd have about 500 pounds of cod. Hake was our mainstay. That's what we've seen decline the most is our hake.

There never was a place for the great big boats and there never will be. The only part of the fishery in Canada that pays its own way, the people pay their taxes, the people provide jobs, is integrated into the whole community and is sustainable year after year is the small, inshore fishery. It doesn't need government help. It can exist on its own. And these are the people who provide jobs for the small plants like you see down at the wharf here. Everything else you see is big government fishery that had to be subsidized, or there were big companies with outside capital that were in for a quick buck, and the only way they could do it was to rape the fishery.

If you are going for 100 per cent efficiency, in about two years or five years at the most, they would catch all the fish that's out there. The shortest amount of time to catch all the fish that's out there would give you the most profit. But then what do you do?

We inshore fishermen are probably not efficient. But we pay our taxes, we educate our children, we do everything and we go on year after year. We just take a percentage out of the fishery like our lobster fishery. It's got a lot of regulations on it. It's our backbone and has been for years. If we had put more regulations on our gear in the hake industry, our hake fishery might be better right today. You can't be 100 per cent efficient and expect to have a fishery every year because if everyone loads their boat every time they go out, there's no fishing.

We were just talking about the tuna. We are dealing with people in Ottawa that are absolutely, totally out of touch with what the way of life is here. Or anywhere for that matter. We deal with seasons and with small changes that make a big difference in our lives. And these people who have a paycheck coming every week make these changes on paper that affect us. They don't realize how they affect us.

We'd like to know when we go fishing tuna. We have quota on tuna that only allow so many to be caught. We don't need any more regulations on the tuna fishery. We have people who will listen to special interest groups to get certain regulations in to get a smaller corner of the market for some particular group of fishermen or some particular buyers that have money or political connections or whatever. That really messes up the works for the ordinary fisherman. These certain people can't understand why we just can't go all the way. These changes they make - they don't realize how much they affect us. A one week change in opening date for certain seasons would be your whole season, whether it's herring, lobster or whatever. It's awful hard to find out about regulations sometimes. It's especially bad on the tuna right now because the bureaucrats in Ottawa made a mistake like bringing in regulations that were set in place for us fishing in Canso. And now they have to save face. So we don't know what they are going to come out with. But it will have to be something that makes them look good. That's the type of thing we're dealing with. We're the ones who suffer.

Whether the ring size in lobster traps is 5 inches or 9 inches doesn't make much difference as far as conservation is concerned. A 4½ inch ring will catch a five pound lobster. A 5 inch ring will probably get a little more. You would probably get the odd breeder in a nine inch ring, but where the trap is that big, most of the canners would swim in and swim out. If you set out to catch canner lobsters, probably a four inch ring is what you should have. There are very few markets, just canners. But it's more of a trap. Where we fish, there are only a few large markets. In an area like Point Graham where they catch mostly markets, they have to use a bigger ring. Canners are from 2 9/16 carapace up to 3¼. They are all processed. Markets are generally sold live. Some people run 50 per cent markets, 50 per cent canners. You have to keep them separated. You don't put bands on the canners, but the markets all should have rubber bands on their claws so they don't bite each other.

Some lobster fishermen might not want to see the cod back, especially if that's all they do. There are other fishermen making a living in the strait, and you have to work with everyone else. There are scallop fishermen and groundfish fishermen out here and they have rights too. We may want to fish cod a few years down the road. It's awful easy to get narrow minded when you're a fisherman.

We belong to the P.E.I. Fishermen's Association. That's part of the Eastern Fishermen's Federation. It helps but we'd like to have something stronger. It always seems like we end up with the short end of the stick on every deal. We see these other big plants that are built in places, and we know they shouldn't have been built; they run five years and they close down. A whole bunch of government money lost. I mean our plants that we fish for, they're all going here. They're all employing people even in hard times. These other plants are closed down and people are out of work because they went into a fishery that would only sustain the boats for four or five years, and they just laid it in and that's it.

A few years ago in Georgetown, the fishery had four or five big boats. Markets go up and down. I'm not saying that a large fishery won't work. It works in Nova Scotia. They've been fishing for years. They do have a place that's a lot better for fish than we have. It's just better waters for fishing in that southwest Nova part. It's the richest part of the Atlantic.

If you're talking about fishing on the Grand Banks and Georges Bank, you're talking large boats there. They have to be. But any fish that's within 25, 40 or 50 miles from the shore can be fished with inshore boats.

The way we've survived over the years is because we all have a lobster license, scallop license, groundfish license and tuna license. Back in the '70s, we were getting three or four thousand pounds of lobster. You couldn't live on that. But the hake fishing was good.

We have a big advantage over the Newfoundlanders, because we have more diversity to fish than they do. There's no lobster on the east coast of Newfoundland. Maybe it has something to do with the temperature of the water. They claim that some years we get two sheds up here in the strait because of the warm water. That really helps the lobster to grow a lot.

Before the offshore factory ships came along, there was ice in the wintertime, and fishermen weren't fishing year around. In the wintertime the fish were bunched up twice as thick as in the summertime. Whereas the fish might be congregated in a five to ten mile area in the wintertime when they're fishing them, in the summertime they would be spread out over an 80 mile area. At first, the government didn't think it was the same fish as the inshore fishermen were catching. They thought they had found new schools of new fish. The consensus is that they are the same fish that migrate. We're dealing with a lot of variables here, boys. There's politics involved. Different companies would go to government with big schemes that looked excellent on paper. Ten years later, it was seen as a total mistake.

The fishermen up here were telling DFO for years that our fish were on the other side of Cape Breton where they went to winter, and DFO said it wasn't so. But just in the last three or four years, they discovered it was true. And that's where their big winter fishery was depleting stocks in the gulf here.

All the time we sent proposals to do with all kinds of fisheries to meetings. It's unbelievable what they will vote down. Total one hundred per cent common sense to do with the fishery. Especially on the tuna fishery where we're dealing with other countries. We have one or two representatives at these large meetings. If groundfish were being discussed, the offshore fishermen's group would have their say too. If it was lobster, it would be inshore only, unless it dealt with the offshore lobster fishery.

I've been fishing since I was 18. We had our problems with fish that weren't so nice in the dayboat fishery. We upgraded our boats, we carried ice, we did all these things over the years. We had to make things pay. The plant employed people filleting fish, doing scallops, processing lobster and processing herring.

Speaking of herring, if you want to be more efficient, you use a seiner. But the only thing is if you seine around herring, and you ignore the laws, what do you do with them? You let them go. How many live? Only a small percentage. You have to seine them up to know what you've got. And once you know what you've got and they are too small, you let them go. It's too late then, they've already smothered. Half of the school might sink to the bottom.

Some people might say that gillnets are not the best way to catch fish. When you use 2 and 5/8 mesh, if your herring is just a speck too small, he swims through. He strikes off a couple of scales and keeps on swimming. He's there for next year to spawn or whatever. But in the seine, if a school is trapped, that's it for that school, whether it's useable or not.

Acronyms

AGAC	Atlantic Groundfish Advisory Committee
APPA	L'Association des Pêcheurs de la Péninsula Acadienne
CFA	Crab Fisherman's Association (PEI)
CNA	Certified Nursing Assistant
DFO	Department of Fisheries (Federal)
EFF	Eastern Fishermen's Federation (NB)
EKFA	Eastern Kings Fishermen's Association (PEI)
FACWC	Foreign Allocations in Canadian Waters Committee
FPI	Fishery Products International
FRCC	Fisheries Resource Conservation Council
GFA	Groundfish Fishermen's Association (PEI)
NATSEA	National Sea Products
NCARP	Northern Cod Adjustment and Recovery Program
NIFA	Newfoundland Inshore Fisheries Association
PEIFA	Prince Edward Island Fishermen's Association
PEIGFA	Prince Edward Island Groundfish Fishermen's Association
TAGS	The Atlantic Groundfish Strategy
UFCW	United Food and Commercial Workers
UMF	United Maritime Fishermen

LOCATION OF THE PEOPLE INTERVIEWED.

Because of the central location of the National Sea Products plant in Louisbourg and the Petty Harbour Fishermen's Producer Co-operative Society in Petty Harbour, the people interviewed there lived at or nearby those two sites.

In New Brunswick the fishers interviewed were scattered along the north shore, and in Prince Edward Island the east coast fishers lived from Souris to Murray Harbour and places in between. Here is a list of the people interviewed in New Brunswick and Prince Edward Island. Opposite the names is the particular community where each person resided.

NEW BRUNSWICK

Leigh Jagoe, Bathurst
Walter Coombs, Bathurst
Harper Smith, Bathurst
Ernie Smith, Bathurst
Frank R. Daley, Janeville
Barry Bezeau, Miscou Centre
Larry Vibert, Miscou Centre
Doug Vibert, Miscou Harbour
Donat Lacroix, Caraquet

PRINCE EDWARD ISLAND

Danny, Jack and Leslie King, Georgetown
Gordie Gotell, Georgetown Royalty
Doug Sorrie, Montague
Edwin McKie, Souris
Bert Boertien, Souris
Eber Williams, Beach Point
Bruce MacNeil, Murray Harbour

MAKE AND BREAK HARBOUR

How still lies the bay in the light Western airs
Which blow from the crimson horizon
Once more we tack home with a dry empty hold
Saving gas with the breezes so fair
She's a kindly Cape Islander, old, but still sound
But so lost in the longliner's shadow
Make and break, and make do, but the fish are so few
That she won't be replaced should she founder

It's so hard not to think of before the big war
When the cod went so cheap but so plenty
Foreign trawlers go by now with long-seeing eyes
Taking all, where we seldom take any
And the young folk don't stay with the fisherman's way
Long ago, they all moved to the cities
And the ones left behind, old and tired, and blind
Can't work for "a pound for a penny"

In Make and Break Harbour the boats are so few
Too many are pulled up and rotten
Most houses stand empty. Old nets hung to dry
Are blown away, lost, and forgotten

I can see the big draggers have stirred up the bay
Leaving lobster traps smashed on the bottom
Can they think it don't pay to respect the old ways
That Make and Break men have not forgotten?
For we still keep our time to the turn of the tide
And this boat that I built with my father
Still lifts to the sky! The one-lunger and I
Still talk like old friends on the water

Fogarty's Cove Music, 23 Hillside Ave. S., Dundas, Ontario L9H 4H7